The Astronomer and the Countess

The Astronomer and the Countess

A Novel of Intrigue at the Forefront of Science

Peter Foukal

HAMILTON BOOKS
AN IMPRINT OF
ROWMAN & LITTLEFIELD
Lanham • Boulder • New York • London

Published by Hamilton Books
An imprint of The Rowman & Littlefield Publishing Group, Inc.
4501 Forbes Boulevard, Suite 200, Lanham, Maryland 20706
www.rowman.com

86-90 Paul Street, London EC2A 4NE, United Kingdom

British Library Cataloguing in Publication Information Available

Library of Congress Cataloging-in-Publication Data Available

ISBN 978-0-7618-7404-1 (pbk. : alk. paper)
ISBN 978-0-7618-7405-8 (electronic)

™ The paper used in this publication meets the minimum requirements of American National Standard for Information Sciences—Permanence of Paper for Printed Library Materials, ANSI/NISO Z39.48-1992.

To Lizzie

"Big advances are not made by analytical procedures but by direct vision. Yes, but how?"

—Lawrence Durrell in *Clea*

Notes and Acknowledgments

The main characters in this novel are fictional. The accounts of scientific facts and historical happenings are intended to be accurate and the author apologizes for any errors. Letters with Czech soft diacritical marks are pronounced as follows: č is ch, š is sh, ž is dj. The soft ř is difficult to transliterate. The Czech long-vowel diacriticals have been omitted for simplicity.

The author is grateful to his wife, Elisabeth Foukal, for her support and careful proofreading of early drafts. He is indebted to Anne Aitken and Brooke Bures for editorial assistance; to Neil and Marybeth Sheeley for suggesting the title; and to his brother, George Foukal, for much sound advice. Finally, he thanks Claire Taittinger for posing the question that inspired this novel.

Chapter One

The news that Jock White, a popular Congressman from Tennessee, had been killed by a cougar while on a hunting trip in Idaho caused a sensation in the national media in the fall of 2015. The coverage described Jock's many contributions to his constituency and gave testimonials to his ability and character from Congressional colleagues as well as friends and family.

The *Washington Post* interviewed the guide who accompanied Jock. "It was a freak accident like I'd never seen," he told them. "That cat just dropped from an overhanging branch when we were riding our horses through a grove of trees and snapped his neck with one bite. Then she gave me a crazy look and a snarl and disappeared into the rocks. I never saw anything like it in my thirty-three years of guiding."

The circumstances surrounding the Congressman's death were investigated by local authorities and ruled to be an unfortunate but natural death. Even in DC political circles, the incident was soon eclipsed by concern over the next year's Presidential election. Only three people—the noted Harvard astronomer Philip Frobisher, the senior NASA administrator Cecily Thomas, and Anna Grad, a talented young intern in Jock White's office—knew that there was more to the incident. They alone knew that Jock's demise was just one facet of a strange matter of some importance.

Chapter Two

A small meeting of a few dozen colleagues assembled from around the world was held in the spring of 2015 to discuss the latest advances in understanding how the universe was born and how it had evolved.

The venue was a lovely château in Normandy not far from Alençon belonging to Bernard, the Count de M. and his wife, Estelle. The Count and Countess were in their mid-forties and made their property available to suitable parties. For years they had enjoyed the company of visitors and the scientists provided a welcome distraction from hunting and other preoccupations of country life. They both also felt it was good for their teen-aged children to broaden their minds by exposure to interesting people from foreign lands.

Philip Frobisher had recently turned sixty and was proud of not looking it. He was a professor at Harvard and was used to deference from his colleagues. He was of somewhat less than average height and not a particularly imposing figure physically but his impatient intensity and forward lean commanded respect. His look said, "Show me that you are worth my time."

Philip had worked hard to get tenure and now enjoyed the benefits like meetings such as this one, in nice places with stimulating company. He hated the huge conferences of the national professional societies held in convention hotels and over-run with people he'd never heard of. His students attended those and reported to him whether anything interesting had happened.

The drive from the train station was long and Philip was a nervous passenger. "God, these French always drive so fast," he thought. He was relieved to arrive intact and looked forward to a drink in the bar. The meeting agenda had listed a reception at six o'clock which gave him a chance to wash and catch up on some messages.

Philip prided himself on having more esthetic sense than is commonly attributed to scientists and as his cab approached the renaissance château along a long avenue of stately poplars he admired the two round towers with conical black tile roofs that provided a touch of whimsy to the otherwise severe design. Although little was yet in flower in late March, the winding green hedges of the formal gardens softened the grey stone walls and the cobblestoned courtyard.

He was received cordially at the entrance by a doorman and taken to his room whose white plaster walls, Philip noticed, were nicely set off by dark ceiling timbers. The antique prints on the walls displayed scenes from 18th-century hunts on the Count's property.

Philip had come to believe that he deserved comfort so he appreciated the private bath. He had made a mental note from previous European visits to be wary of the new custom of showers that flowed out into the whole bathroom. The pretty tiled floors were as slippery as a skating rink when wet.

"Wonderful to see you, Philip," he heard as he emerged into the hall a little later. It was the booming voice of Nigel Worthington from Cambridge University, a colleague of long standing whose shaggy beard and garrulous manner belied a keen mind. "I heard that your people have observed some curious behavior in M 87," Nigel continued as they shook hands. "Ah, yes," answered Philip distantly as they walked towards the reception together. He was still recovering from jet lag and didn't want to start sparring with Nigel until he'd had a drink or two. He'd come to realize that, even in discussions of the universe, form was as important as content.

They joined a few others assembled in the lovely baroque salon. The ornate style with its powder blue walls, gold gilt mirrors and elaborately veneered furniture was not to Philip's taste, but he appreciated that this room was not a sterile showplace; a certain level of disorder betrayed that the family lived here.

Professor Zekhulin, a clever young theorist from Moscow University, strolled over and greeted him cordially. "What is new at my dear Harvard?" he asked. "Parking is as challenging as ever," responded Philip curtly, looking around for an escape route. Zekhulin was ambitious and angling for an invitation to bolster his resume but Philip didn't care for him enough to be bothered with all the visa formalities.

Dr. Nanda from the Tata Institute headed a group that had recently published a pioneering set of radio frequency observations of black hole environments. Philip had visited them in Pune not long ago and was grateful for their warm hospitality. He remembered a conversation with Nanda over tea on the institute's pleasant verandah one afternoon. "Arvind," he had asked, "do you feel justified in spending vast sums on new instruments when hundreds of millions of your countrymen are starving?" Nanda had clearly fielded that question before. "We in India cannot wait for everyone to be well fed to begin doing research. We all understand that," he had answered, smiling.

Others whom Philip recognized included Sandra Wells from Stanford University and Bob Little from Caltech. Their joint study of the microwave background was helping to explain the evolution of the universe's expansion and the reason why it seemed to be accelerating rather than slowing down. Philip was more at ease with observational astronomers like them, than with the theorists. He understood the theoretical explanations well enough to identify the most interesting new measurements to make. But he couldn't follow them down their mathematical wormholes into the Alice in Wonderland world they had constructed for their own delight.

"Hello, Phil, are you ready to puzzle over Zekhulin's latest brainchild?" joked Sandra. He had confided to her once that, although he had taught a section on cosmology to Princeton undergrads in the 1980s, he didn't pretend to understand what all the new stuff using string theory was about. Back then, when he was a postdoc, you could explain the universe's evolution to students using simple models no more abstruse than the physics of a falling apple. Now, observational testing of current developments in string theory was impossible and some even argued that it was not necessary.

"I'm looking forward to dinner," he confided with a grin to Sandra and Bob, nudging them towards the dining room lit up with sparkling candelabra reflecting from the floor-to-ceiling French windows. Sandra was recently divorced with several children; and meetings gave her a chance to kick up her heels. She was pert and personable but not Philip's type. Besides, he was happily married and had no interest in any hanky-panky. He preferred not to be seated next to her at dinner.

Philip soon discovered that most of his colleagues had found places more quickly and he was left looking around for a free spot. Finding none next to anyone familiar, he took the last chair and focused on the heavy silver cutlery arrayed before him. As he sat down his peripheral vision took in a strikingly beautiful woman sitting across from him at the long dining table. Judging by her perfect hair and confident presence, he divined that she was probably the Countess.

This both stirred Philip and annoyed him. He was tired from his trip, not at his best, and now found all eyes on him, expected to proffer witticisms and amuse their disturbingly lovely and expectant hostess. He introduced himself with "bonsoir, Madame, my name is Philip Frobisher," but she only smiled in amusement, looking at the name tag displayed on his jacket. Not a great start, he thought. Asking about her children, usually guaranteed to get people talking in Boston, was no more successful. "They are studying for their examinations," she answered. She was clearly not obsessed with them.

After some awkward silence she looked at him intently with her shockingly blue eyes and asked: "Tell me, Mr. Philip from Harvard, what do you do exactly?" Philip took a while to recover from the directness of this rather simple question. How should he explain to a lovely French Countess, over blanquette de veau, what he did? Worse yet, the whole table had heard the question and was dying to hear his answer.

"Well, Madame, have you ever visited an observatory?"

"A long time ago, with my parents, we visited the telescopes on the Pic du Midi in the Pyrenées where a strange relative was working," she offered.

"Excellent," he continued, "I travel to observatories like the one you visited making measurements of objects like stars in the sky, using the biggest telescopes in the world. Some are much bigger even than this château," he added, for local color.

"But not as old and beautiful," she pointed out, with a ravishing smile.

"Certainly not," agreed Philip, thinking that he had never heard English spoken with such an absolutely charming accent. He relaxed and looked up and down the table to enjoy how well he had acquitted himself. But the Countess was not finished. Once again fixing Philip with a sweet look, she asked: "To become a professor at Harvard you must have discovered something important. What important thing have you discovered?"

Good grief, thought Philip, this is worse than my Ph.D. oral.

"You know, Madame, that is an excellent question," he responded smiling. "It is a question that scientists should be asked more often. Let me give it the thought it deserves and answer you at dinner tomorrow."

Conversation then turned to the Countess's interests in hunting and to her horses. After the local cheeses had been tried and pêche melba and coffee had rounded off the meal, the company drifted off to bed in the former servants' quarters of the château. Philip had trouble falling asleep even though he was very tired. He was happily married but intrigued.

Chapter Three

The next day was taken up with scientific talks and a panel discussion. The meeting was in part a visioning exercise aimed at gathering the most senior scientists in the field to chart the direction of the future. Philip chaired the panel that had been formed for this exercise, a function that required leadership and diplomacy. Scientists are the most status conscious element of society and no vision for the future will gain acceptance unless respect is paid to their intricate pecking order.

Philip knew that he was chosen as chairman to lend credence to the report that the panel was expected to send to the funding agencies, NASA and the European Space Agency (ESA), who were paying for the meeting. He had come to accept that his Harvard affiliation, more than any personal accomplishments, led the agencies to put him in charge of an important planning exercise like this. He was past any sensitivity on the topic; he'd put everything on the line for a decade to get tenure, and had maneuvered skillfully during his years as a graduate student at Caltech and postdoc at Princeton to get the tenure track faculty job at Harvard. He saw no need to feel apologetic.

Still the Countess's question lay in his mind. What *had* he discovered that was important?

The panel discussion went well. The preparations he had made, initiating discussion among the half dozen panelists by email beforehand, bore fruit and the ideas flowed smoothly. Thoughtful answers to likely questions from the floor were in place. The agency representatives were impressed and when they broke for lunch in the château garden, they hastened to congratulate him and the other panelists.

Philip looked around for the Countess but she was nowhere to be found. He was disappointed because he was looking forward to discussing her question. But more than that, he was hoping to see where the conversation with her would lead. The food did provide some consolation. Cold and warm soups, delightful little canapés, a marvelous assortment of cheeses and a pastry cart that demanded Philip's full attention. Still, while he sipped his coffee, he still hoped that she might appear.

He sat for lunch with several colleagues and discussed the main theme emerging from the discussion: Many cosmologists felt that a space-borne telescope twice as large as the largest now under construction, the James Webb Space Telescope (JWST), would be needed to advance their field. It was not a very creative goal but it made NASA and ESA happy.

As usually happens in such visioning exercises, the agencies had pretty much decided beforehand what they wanted to emerge from the report. The general path of the scientists' endeavor was guided, some would say dictated, by the need to provide employment in the US aerospace industry. The Europeans, and now the Chinese and Indians, were operating along similar lines in their own national programs. As long as all the sub-systems were equitably distributed amongst the participating nations and aerospace firms, everyone was happy.

When science had consisted of a few pipe-smoking, tweedy types often willing to pay for lab equipment out of their slim salaries, they were welcome to do as they pleased. But those days were long gone. Now, governments not only needed science to further national goals in commerce and defense, they saw in it an important way to keep their multitudes busy. For this it mattered less that anything important be discovered. With our planet's population approaching eight billion, the imperative was to keep people employed in plausibly useful and satisfying activities. Few questioned whether spending heavily to achieve this aim was worthwhile.

Scientists seldom discussed this kind of behind-the-scenes stuff; they were happy to be given big budgets that enabled universities and national labs to grow. They didn't question studies which confidently asserted that big groups produced the best science and that the quality of science increased with increased funding. No one who thought about these assertions could really demonstrate that they were true. But more funding certainly supported more scientists and for most, that was really all that mattered.

In the afternoon session the conference broke up into a few working groups. Philip joined the one dealing with the most difficult issues of organization and funding of such a vast project. Would the US go it alone or seek foreign partnership? Nowadays, much of the money might come from philanthropists as well as NASA.

Still, if the JWST over-runs were already taking that project's cost over the $10 billion mark, the price tag on its successor would be staggering. Previous experience with the biggest projects attempted by scientists, the Superconducting Super Collider (SSC) of the 1980s and the ITER fusion machine limping along at present, suggested that there was danger in committing a field's resources to one huge project. If over-runs forced termination, as had happened with the SSC, the field might implode. Energetic particle physics had been the flagship of the physical sciences between the 1950s and '70s, but when the plug was pulled on the SSC in 1993, hundreds of particle physicists left the field for astronomy and for Wall Street. Thirty years later, US particle physics had still not recovered.

Philip's group wrestled with the options all afternoon and was pleased to be reminded that it was time to call it a day. It was a glorious spring evening and with the warming of recent years, the lilacs were already coming into bloom filling the air with their sweet perfume.

Philip was happy to step outside and still enjoy the sunshine. He saw the Countess's children walking their horses to the stables across the courtyard and he strolled over to chat, asking whether their mother was home.

"Oui, Monsieur," responded the daughter: "Maman was only in Paris for the day; in fact, here she is now."

Philip turned to see the Countess, lovely even in country dungarees, walking towards them across the cobblestones. She embraced her children and shook hands demurely with Philip, introducing her daughter and son to him.

"I heard that you were in Paris, was it for shopping?" Philip asked with a smile.

"No, not at all," she responded quietly. "I went to visit an elderly man who works for our family, he is in the hospital."

Seeing his surprise she continued, "You know, when you or I must stay in the hospital we understand the surroundings and we can keep busy with a book or a computer. But for a simple man like him, it is frightening. He does not know what to do except to stare at the wall. I went to reassure him." This matter of fact explanation of her trip to Paris surprised Philip; such a sense of noblesse oblige was outside the range of his experience.

"Would you like to see the grounds of the château?" asked the Countess. "We could go for a little walk with my dogs before dinner." "I'd like that very much," responded Philip eagerly, convincing himself easily that life is short and his colleagues could think what they liked.

They set off with the Countess's two Labradors, she in her Wellington boots and he trying to navigate the muddy track in street shoes. They crossed the formal garden behind the château along a pebbled path and continued into the woods. "We own a large property and I manage the farming and forestry as well as our horses," Estelle explained.

"Your husband leaves it all to you?" asked Philip. "Mainly, yes. Bernard is a banker and he is away a lot," she responded. "It is a great privilege to own such a property, but it also carries responsibility. Both to the land and animals and also to the people who have worked here all their lives, and even sometimes their parents and grandparents too. Sometimes we wish that, instead of inheriting it, we could have chosen a smaller place just our own."

Philip didn't know anyone back home with such responsibilities. He had met a few members of the venerable Boston families but their young generation seemed to live pretty much as they pleased. They were happy to serve on boards of hospitals and schools, but the days when they controlled the city's commerce and government were a distant memory. They had been largely supplanted in those functions by immigrants from Ireland like the Kennedys and then later by waves of enterprising newcomers from Italy, Greece, and most recently, Jews from Russia.

"What does your wife do when you are away?" she asked. Philip was taken aback that she should take interest in his home life, but managed to answer: "Our daughter is applying to university and our son to a boarding school. In America parents like us spend a lot of effort trying to get their children into the most prestigious and selective schools, so much of her time is devoted to that activity these days."

"Yes, I have heard about that," she answered, at the same time calling to her dogs who were racing through the forest in pursuit of some fleet-footed prey.

After a half mile they arrived at a large pond in the woods, almost a lake. The road they were following crossed along a causeway fringed with an early show of daisies and other spring flowers. From there they could admire the view across the water to a meadow extending westward. The dogs tore into the water and emerged soaked. When they both ran to him to shake off, soaking his pants, she laughed cheerfully: "They like you, a good sign!"

She then turned to face the setting sun and Philip sensed a sudden sadness that he did not want to intrude upon. "This was the favorite place of our eldest boy who died two years ago," she volunteered quietly.

"I'm very sorry," said Philip. "How did he die?"

"He suicided himself. We could not understand why. He was talented and seemed happy, with a promising future. I am still thinking all the time why it happened. Afterwards," she added, "my husband spent less time here with us. He says that the memories are too painful for him."

Without thinking, Philip reached out and took her hand. It felt the natural thing to do. They stood together looking out over the water in silence. He wondered how it was that he felt closer to her, whom he had only met the day before, than to many of their friends back home. The relationships there often felt more like a competitive tug-of-war. But he sensed by the inflections of Estelle's voice and the engaged look in her eyes that she sincerely cared about what passed between them.

They walked slowly back. The lights were already sparkling in the windows and the guests had moved well into apéritifs as they approached. It was a cheering sight. "Make yourself ready for an excitement this evening," she said teasingly in her charming English. "I will play some piano for your group after dinner, and my daughter, Eloise, will sing."

Chapter Four

Philip joined his colleagues for a quick glass before dinner. He drank Scotch all winter but preferred Pernod when the weather turned warm although it brought back memories of an episode he had come to regret. A few years after he and his wife, Emily, were married, he was offered a fellowship to spend a few months at the Nice Observatory on the Côte d'Azur. Sadly, in retrospect, they felt that they were too busy and passed it up. A more adventurous colleague had taken advantage of the opportunity.

When the colleague returned he described how he and his young wife had rented a studio in a Victorian villa perched on a cliff off the Moyenne Corniche, the dramatic coast road between Nice and Monaco. Their little deck, he went on, looked out over Cap Ferrat and the sparkling Mediterranean stretching towards Africa. They used to sit out there, he recounted enthusiastically, with their feet up on the railing, watching the sunset over the sea and after dark, they were delighted by the twinkling lights of the little fishing boats in the bay.

Philip later wished that he and Emily had shared times like that before the kids came. He remembered how that old roué, the Duke of Windsor, had once remarked: "It isn't important to be in love all your life, but it is important to have loved." Philip thought wistfully that the Riviera's red cliffs, perfumed breezes and green-blue sea might perhaps have helped him and Emily make a better start on the Duke's maxim.

Now, as Philip tried to focus on the organization and funding of the new telescope, he was having difficulty in following what Nigel was telling him. He'd never been adept at eating passed hors d'oeuvres and balancing a drink while talking. Looking around trying to spot the

Countess made the task even trickier. But Nigel was not to be deterred. Speaking loudly enough so others could hear, he proclaimed that the British were unlikely to join in the project because of rising concerns that astronomers should study data from large telescopes that had only been in operation for a few years before rushing into building a hugely expensive new instrument.

The French and Italians felt similarly. In any case they and the Germans had, in the past, mainly joined with other European nations in building their own projects. They had been disappointed too often by cancellations of joint missions with NASA, forced by year-to-year changes in Congress's budgets for US space science.

Philip had to admit that these were valid issues. But he pointed out that the monster ITER fusion energy project underway in the south of France showed the possibility of a truly global collaboration if the goal was sufficiently persuasive. That brought derisive laughter. "No, on the contrary," exclaimed Gerhard Schlitz from Potsdam, "the mess at ITER and previously at the SSC confirms that scientists cannot be expected to manage projects of that huge size."

The conversation was interrupted by the call to dinner. The Countess arrived late and seated herself with another group, so Philip took comfort in the chef's successful langue de boeuf, sauce Madère, and the excellent accompanying Bordeaux. He tried to focus on the broken English of two Japanese colleagues over dinner, but was distracted by concern over the direction that the conference discussion was taking.

The US agency representatives in particular were highly attuned to what they were hearing about the project. As chairman he had to guide the deliberations and must not lose their confidence. It would be bad for his reputation if this conference unraveled and they all went home without momentum behind the project. For NASA there was too much at stake and failure was not an option. He'd need all his experience and diplomatic skills tomorrow to make sure that they could report good progress to Washington.

During dessert the Countess rose and the room hushed. "I welcome your distinguished company to our home and family," she began. "My husband, Bernard, regrets that he cannot be present, but as an amateur of astronomy himself he takes a keen interest in the outcome of your discussions."

Then she continued, "After dinner, I propose to you a short musical entertainment in the salon; I will play a well-known piano piece by Beethoven and afterwards I will accompany our daughter, Eloise, who will present a few songs by Schubert."

The company moved to the salon and the lovely music evoked for Philip scenes from Victorian novels of social evenings featuring impromptu performances by the hosts and guests. He wished there could be more of that these days. The Countess played even the difficult third movement of the "Moonlight Sonata" very well and Eloise then sang a couple of Schubert lieder including the lovely "Auf dem Wasser zu Singen." She was not as striking as her mother but looked pretty with a plain barrette in her hair, radiating a calm that was remarkable for a girl of fifteen.

It was a memorable end to a fine day and all parted in good humor. Afterward, Philip sought out the Countess and Eloise and congratulated them on their performance. "Would you like to have a little visit of the château tomorrow and meet my horses?" asked the Countess, somewhat diffidently, he thought. "Of course," he answered. "We have a short work day tomorrow so the group can tour some nearby historic sites. I've visited Caen and the Normandy beaches before so I'll stay here to do some work, and I'm grateful for your offer."

Chapter Five

Dave Powers had not been invited to the meeting. He was a program manager at Ball Aerospace, a respected space hardware firm in Boulder, Colorado. But it was his job to track big space science projects like this, so he had managed to arrange permission to attend as an observer. It was somewhat of a grey area whether his presence violated contracting regulations, but NASA was fairly liberal in its interpretations and no one expected trouble since the discussions would all be available online anyway.

Dave's father had been a renowned atmospheric physicist, so he was used to the scientific community and had even inherited some credibility with a few of the US astronomers whose instruments had been built in part by Ball. He was better acquainted, though, with the NASA Astronomy Program Director, Paul Markarian, and his boss, the Space Science Division Director, Cecily Thomas. Both of them were present, trying to shepherd this mission through its perilous early phase.

He joined them at breakfast the next day, which was always a tricky gambit. "How are you managing in French, Dave?" asked Cecily, trying to be supportive. Paul just stared into his copy of the *London Times*. Some people were able to manage a fork and knife and converse coherently early in the day, even when jet lagged. Others were just morose and resented company, you never knew.

"Luckily, I haven't needed French much here yet. Do you speak the language?" Dave answered, giving her an opening to show off her broader talents. "I manage," she answered, injecting some calculated self-deprecation. "I spent a year abroad in Montpellier as an undergrad; I really like the French."

“What do you think of how the meeting is going?” ventured Dave bravely.

“The science seems to have been laid out pretty persuasively, don’t you think?” she tossed back to him, daintily wiping some egg from the corner of her mouth.

Dave wasn’t so sure. “As I see it, the main drift seems to be that bigger is sure to open up new, unanticipated horizons, even if there are few specific needs for a larger instrument based on what we now know,” he offered, while hurriedly pouring himself coffee and buttering some toast.

“Well,” answered Cecily, “I sat in recently on discussions at the National Science Foundation regarding the next–generation energetic particle accelerator. There are similar concerns there. CERN bet big on the Large Hadron Collider in Geneva and except for the Higgs boson it really hasn’t produced anything significant. The promised new frontier hasn’t materialized.”

Paul had finished his almond croissant and looked up from his *Times*, checking his watch. “Time to head off to the first session,” he muttered. He was annoyed that Dave had intruded on his time with Cecily. It was hard to get her attention and he needed to ask for a favor.

Program Directors like he at NASA and the NSF were squeezed between the needs of the scientists who depended upon them for funding, and the aspirations of their agency’s upper echelons. It was a tricky balance. Scientists generally thought that they were at their Program Director’s mercy, but it wasn’t that simple. Agencies benefited whenever one of their programs funded a renowned researcher. It gave the agency the visibility needed to get more funding from Congress. So, funding a Nobelist could be good for a Program Director’s career. It was more of a symbiotic relationship than many scientists realized.

Recently, however, Paul’s life had been complicated by a development that he hadn’t discussed with Cecily yet. A well-known scientist whom he was funding had become embroiled in a dispute over ethics. The matter was made trickier because the dispute was between two female senior researchers. In today’s climate of gender sensitivity, he preferred to stay clear of this potentially damaging imbroglio. He was hoping that Cecily would resolve the dispute. "Maybe I'll manage to sit with her this afternoon on the excursion bus to Caen," thought Paul.

As the morning session began, Paul expected to hear the familiar arguments that Philip used to justify increased funding: funding of his group at Harvard, of various projects that he had been associated with, and of the field of astronomy in general. He had been listening to Philip since they had been graduate students together at Caltech thirty-five years ago, so he didn't expect to learn much that hadn't been said before.

He had nothing against Philip personally; he was a decent guy as scientists go. Nevertheless, he couldn't avoid being annoyed. They had started off together but now Philip was an influential professor at Harvard, while Paul's NASA career had stalled. Philip hadn't been much better at the graduate courses they took together. But he made a better choice of Ph.D. advisor; someone who was influential and could help his career. Philip had also worked the meetings that they had both attended in those days, making friends with senior people who could help him on his way.

Paul didn't wake up to that kind of stuff until it was too late. He had to settle for a post doc in the sticks while Philip went to Princeton. Perhaps most important, Paul never seemed able to identify his own topic, his own path. As a postdoc, he worked in a large group analyzing someone else's data. Meanwhile Philip was already writing his own proposals with a supportive faculty member.

Philip made a smooth transition to tenure track at Harvard while Paul couldn't find any university faculty employment. He was grateful for a job at NASA reviewing proposals. Their careers continued to diverge and twenty-five years later, here they were. Philip was always friendly to Paul, but they didn't socialize much besides asking about one another's families. Really, they lived on different planets.

Chapter Six

Dave seated himself at the very back of the meeting room and unpacked his laptop. If proceedings dragged, he was thinking, he could keep in touch with his team back in Boulder. But when he gave presentations himself he hated to see half the audience staring at their screens, so he closed the laptop when Philip entered the room to take his place as chairman of the panel.

Today was the last day of the meeting so Philip started with a brief overview of the arguments made so far. Dave was preparing for a relaxing morning. Then, Philip switched gears. Walking around the lectern, he sat down on the edge of the long table in front of it, to get closer and more personal with his audience.

"Let me tell you a little story to make the point I want to put before you," he said. "When I was a grad student at Caltech the person I admired most was Bob Leighton. The world-renowned theorist Richard Feynman considered Bob to be America's finest experimental physicist. In the 1950s Leighton got interested in what could be measured on the sun with an old tower telescope on Mt. Wilson, overlooking Pasadena and Caltech.

"While other accomplished nuclear and particle physicists were working with fancy new accelerators, Bob went his own way and developed some amazingly sensitive techniques for measuring motions in the sun's atmosphere. Just sort of optimizing some old-fashioned photographic techniques, believe it or not, in a dark room up on Mt. Wilson. Using those techniques he and his grad students discovered that the sun oscillates with a five minute period, which opened up a whole new field of helio- and astero-seismology that has enabled us to look into the interiors of the sun and stars, measuring their temperature structure and internal rotation.

"He also discovered the solar supergranulation, a previously unknown scale of motion in the sun's atmosphere, that led to advances in solar dynamo theory, the basis of our understanding of the magnetism of the sun and similar stars. In five years Leighton advanced understanding of the Sun more than anyone else had in the previous fifty.

"Then Bob switched fields completely. He got interested in millimeter wave astronomy and personally oversaw the building of Caltech's millimeter array that obtained pioneering observations of the most obscured regions of our galaxy and of star formation.

"Why am I telling you this? When I was leaving Caltech for Princeton, my last stop was a visit to Bob. I wanted to tell him how much I admired his achievements. And I wanted to ask him why he had moved on from brilliant success in individual research to take up management of a big project in a totally different field."

"Philip," he answered, "there are different stages in a research career. When I was young, individual, small-scale research was where I could make most impact. Now that I'm older, with some important achievements behind me, people listen to what I have to say. Now I can contribute more effectively by putting my influence behind a big project that a younger man couldn't pull off."

"I remembered what Bob told me," Philip continued, "when our perceptive hostess asked me at dinner on our first day here, what I had done that was important. Quite frankly, after considering her question for the last two days, I can't say that I've done anything really important. I mean important enough that it would be mentioned in an introductory textbook on astronomy.

"Oh sure, I've developed some faster techniques for measuring galaxy red-shifts, and managed a sky survey that improved our statistics on the expansion of the universe," he went on. "But nothing I can really call innovative and ground breaking like Leighton's findings with a rusty fifty-year-old dinosaur of a telescope.

"So I'm afraid that devoting myself to this project would consume the rest of my career, and I don't want to do that without trying still to do something really innovative and impactful." Looking around the room with a sheepish smile, he added, "Something important enough that our charming hostess would agree would justify my appointment as a Harvard professor."

No one had expected such a speech from Philip and consternation reigned. Cecily wondered how she could keep this train on the rails. But she took over the lectern after Philip sat down and put into action the skills as conciliator that had earned her rapid promotion at NASA. Thanking Philip for a most honest and interesting commentary, she asked whether there were any questions. Hearing none, she suggested that they all break for coffee.

Chapter Seven

"What was all that about—a case of mid-life crisis?" asked Nigel. He had sought out Cecily at coffee break. She was also English so he felt that there was less danger that exchanged pithy allusions might go awry. "I suppose I should start looking for someone else to lead the charge," she responded. "It may be from a quite different field, like investigation of extra-solar planets."

"I see your point," answered Nigel. "We cosmologists don't seem to carry the weight we once did. Nothing really important has come up since that American work on the universe's accelerating expansion." "Yes," added Cecily, "and the field's image wasn't helped by the awkward retraction a few years ago of the results on the inhomogeneities in the microwave background.

"The search for life on other planets really fires the imagination of Congress and the taxpayers," she offered. "We were definitely going to involve that community in selling this project, but now we may need to put them front row center."

"It's ironic, you know," mused Nigel, "about a century ago a relatively unknown observer named Vesto Slipher at Lowell Observatory in Arizona took some spectrograms of nebulae that first showed the systematic red shift versus distance relation of an expanding universe. He used the telescope that the wealthy Boston amateur Percival Lowell had installed to study canals and intelligent life on Mars. The canals turned out to be an optical illusion but Slipher's results launched modern cosmology. So your project may not be the first time that cosmology has piggy-backed on the search for extra-terrestrial life."

"Ha, ha, very clever, Nigel," laughed Cecily. "I'll have to bring that up with my boss when I get back. He'll enjoy that."

Dave Powers wandered into the coffee area, troubled by what he had heard. He would have liked to call back to Ball Brothers to discuss this unexpected development but they were eight hours behind and he wouldn't be able to reach them until everyone here had returned from the afternoon excursion. By then, the participants would be dispersing to catch trains or rides back to Paris and return home. It would be too late to collar Cecily or someone else who might have some useful insight into what was likely to happen next.

Dave was used to the fits and starts of Federal contracting and he doubted whether postponement of this big project would threaten his job. As a Program Manager he had seniority, after all. But it might affect the raise he was sort of counting on if Ball got a study contract for at least one of the payload instruments that would be fed light gathered by the main telescope.

Ball wasn't big enough to be the prime contractor on a job of this size—that would probably go to Lockheed, Northrop, or Martin Marietta. But even the sub-system study contracts could be substantial and lead into the more lucrative hardware phases that follow.

Dave and his wife had just bought a bigger house outside of Boulder in anticipation of a third child and their plan was that she'd quit working at that point and stay home with the kids. But Dave needed that raise to make all this happen, so the turmoil this morning made him nervous.

As the coffee break wound down Cecily once again took charge and announced that, given the morning's developments, the group would leave for the excursion a bit earlier than originally planned, so would everyone kindly fetch comfortable footwear and jackets in case the weather changed. "We will stop for a nice lunch on the way," she reassured them, knowing that armies travelled on their stomachs, especially when several gastronomically well-known country inns lay along their route.

Paul hovered around Cecily, hoping to time it right and grab a seat next to her on the bus and ask his favor. But she was deep in conversation with her equivalent at ESA and kept turning her back to him. As the bus filled Paul found himself sitting next to Dave, whose tales of Colorado ski exploits only reminded Paul that he'd failed gym in high school. As the bus rolled on through the picturesque Normandy countryside, Paul stared out the window, day-dreaming of early retirement and maybe buying a little marina over in Virginia. Over the years he'd gradually come to enjoy boating more than astronomy.

Philip was aware that he had caused a lot of drama but he didn't let it get to him. He didn't feel bound to any particular research agenda and dropping out of this project seemed the right thing. "Where do you plan to hunt for a really important alternative project?" asked Sandra. "No idea yet," he answered. "But look at a friend of mine who was given early tenure at Caltech in the expectation that he would continue their program in solar astronomy. Instead, he surprised everyone by developing an interest in snowflakes and became a world expert on their structure."

Sandra acknowledged that the attribute that top universities were all competing for was the best minds. What those minds did afterwards was of less interest. Of course there were well known misfires like a boffin who was heralded as a genius slated to do great things in astrophysics. After he was made the youngest tenured professor at one of the Ivy League universities already in his late twenties, he went on to produce most notably, a paper on the fluid dynamics of waterbeds.

"The art of finding great minds that don't fizzle is still a work in progress," Sandra freely admitted. "In fact," she added, "what even constitutes a great mind is a conversation for another day."

But Philip's achievements already put him beyond that threshold of becoming an embarrassment to Harvard. He was confident that his little demi-tour here in France would be counted as a mark of eccentric genius back home. Now he was looking forward to meeting the Countess for lunch and for a visit around the château.

Chapter Eight

The Countess had invited Philip to meet her in the foyer around noon, saying she would prepare some lunch for him and Eloise. She was giving the kitchen staff the morning off since the group was leaving on their excursion.

Soon after he had settled himself to wait for her, Philip heard the rumble of an expensive car exhaust and the crunch of tires on the gravel drive. Looking out of a window he saw the Countess stepping out of a lovely red Alfa Romeo two-seater, carrying a bag of groceries. She swept through the door, fetchingly removing her pretty headscarf and tossing her hair to sort out some stray wisps blown by the wind.

"What a great car!" commented Philip admiringly. "Yes; a present from Bernard to give me courage after Thierry died," she said wistfully. "But it is much too fast, even dangerous in the winter. Come with me, we are eating in the kitchen," she added. Philip took the groceries and followed her down a corridor. When he commented on the modern facilities in the ancient château, she smiled and asserted with emphasis: "I love old houses but the kitchen and bathrooms must be modern!"

"I have found some of the first wild mushrooms of the season in the marché this morning," she announced happily in the kitchen. "So I will make a little omelette aux champignons and a nice salad. Also I have bought my two favorite local cheeses for you to try. Perhaps you could open this bottle of red wine while I cook. Eloise will join us soon, she will be happy to make a good salad."

Philip was glad to be useful and managed to open the bottle without destroying the cork. He had never been confident using the classic French waiter's folding corkscrew that yanked the cork at an angle instead of straight up. But today was a lucky day so decorum was preserved.

He poured a glass for each of them. "To the chef!" he proposed, raising his glass, but avoided proffering it, uncertain whether he should expect her to clink with him. Perhaps, he thought, a Countess might consider that a gauche American habit.

"I am not a good cook," she apologized. "When I was growing up we had a wonderful cook who spoiled my sister and me, so I have only learned to make biftek and omelettes," she laughed. Philip thought that he was willing to overlook this deficiency—just sitting on a high stool and watching her move around the kitchen was enough for him.

Eloise entered and greeted Philip cheerfully, offering one cheek to kiss and then the other. "Chérie, please make us a little salad; there are fresh tomatoes and lettuce and some radishes on the counter," requested her mother off-handedly.

Lunch was very pleasant. The simple food was excellent and Eloise talked excitedly of their scheduled trip with her mother to buy some clothes in London. "Aren't the best women's fashions in Paris?" asked Philip. "Of course," answered Eloise, "but nice woolens are cheaper at Marks and Spencer."

Philip asked her innocently why clothes seemed to look better on French girls. "Oh, we wear the same as British girls but we buy them a size smaller," she volunteered sweetly. Philip couldn't imagine his daughter or any of her Boston friends sharing such confidences with a stranger back home.

"Let's clean up, and I'll take you around the château," proposed the Countess. "Eloise can join us for a while to show you her horse." All this was new territory to Philip but he enjoyed being taken into the confidence of these two charming women, so he looked forward to seeing where else this afternoon would lead.

They crossed the cobblestone courtyard and walked down the line of stalls past some that held horses kept here by friends or by other members of the local hunt. Bernard, Philip was told, had been master of the local hunt back when he spent more time at the château.

Philip assumed that they meant fox hunting, which he had seen once at the Myopia Hunt Club near Boston. "Non, non!" exclaimed the Countess emphatically. "Here in Normandy we hunt the stag through the forest, not the fox. The English also used to hunt stags centuries ago, but they lost their forests to ship building and had to change to hunting foxes across open land. It is quite a different sport."

"You should come to visit us on the fête du St. Hubert in November," said Eloise, smiling at Philip. "It is a great event to celebrate the patron saint of hunters and Maman looks especially fine in her French hunting blue," she added, looking at him indolently while stroking her horse's muzzle.

"You are kind to offer, but what would I do, since I've never ridden a horse?" asked Philip. "I assure you that there is much else for you to enjoy," offered the Countess, giving Philip a winsome smile.

Chapter Nine

The Countess reminded Eloise of her approaching examinations and she excused herself, bidding goodbye to Philip in case he left before she would see him again. Philip asked the Countess whether he was keeping her from her duties, but she dismissed that idea with a wave and invited him to follow her. "I am proud of this old house," she assured him.

They walked along a corridor hung with family portraits and Philip leaned over to read some of the names and dates. "I see some from as far back as the 16th century," he commented admiringly. "The valuable ones have long since been sold off when Bernard's family needed money to repair the roof," she laughed. "What is left is of lesser quality and really needs restoration," she added.

Her husband, she explained, was descended from one of the original Norman families whose ancestors were the Vikings. They attacked this part of Europe in the tenth century and some of them settled in the area. His forebears fought under William the Conqueror at the battle of Hastings in 1066, when the Normans also became rulers of England. An original castle dated from that time but it was destroyed and this château was built on the site.

"So, in principle, we and English aristocracy are relatives, but the relationship has never been happy. The northern French have never gotten over the devastation of the Hundred Years' War, and there has been constant conflict with the English that continues even now."

"You say the northern French—are English relations better with the South of France?" asked Philip. "Curiously, they are," she responded. "The Bordeaux region and that whole area called Aquitaine has always enjoyed warmer relations with England because of the wine trade and because their Eleanor was such a popular English Queen. They also share the rugby culture with the English, Welsh, and Scottish. It is all very interesting if you like history."

She led Philip into a library paneled in dark wood furnished with well-worn leather armchairs and sofas. "This is Bernard's favorite room," she said wistfully. "It seems very empty now." Philip looked over the leather-bound volumes lining the shelves and the ancient celestial globe depicting the fanciful creatures of the constellations.

How hard it must be, he reflected, for Bernard to be deprived of a life that connected him to his family going back a thousand years. How quickly their life was changed by the death of their son. No one could ever be sure what lay ahead. Neither in their personal life, nor, for Philip, in his career.

He followed the Countess up a grand staircase. "Eloise looks forward to descending this staircase in her wedding dress one day," she laughed. "I hope she gets her wish. The French of her generation are marrying less and less. Perhaps she will find a foreigner who will make her happy," she added jokingly.

At the top of the stairway a door was open into a large, airy room furnished in light pastel colors with wallpaper showing cherubs, nymphs, and satyrs in the style favored by Boucher and Fragonard and other painters of their time. A delicate lady's desk inlaid in rare woods was set near a large window to catch the afternoon light. "This is my favorite room, where I write my letters and read in the evening. I handle the bills and other not so pleasant business from an office downstairs," she told Philip.

"Come, let's sit here for a while. I'll ring downstairs for some coffee and perhaps a little pâtisserie if the maid has returned from the marché already. You must, after all, tell me how your meeting went. I am going to Paris tomorrow to spend a few days with Bernard, and he will ask me scientific questions, I am sure!"

Philip smiled and did his best to get comfortable on one of the overly stuffed formal chairs, wondering to himself how he should respond. "The big telescope we came here to discuss will, I am sure, go ahead.

It would be un-American not to reach for the next bigger thing. But I have decided to let someone else lead the effort," he explained. "I prefer to devote myself to work that may lead me to findings that are truly important and really mine."

"So that is your answer to my little question, that you believe that your important work still lies ahead of you?" "Yes, Countess, that is my answer," he replied, looking at her fixedly. "Mon Dieu," she exclaimed, "I hope that my silly question did not cause this trouble for you!"

"Your question wasn't silly at all. It just made me realize that time is running short for me to make an important and original impact."

"But what will they say at Harvard?" she asked. "Harvard and other top universities prefer their professors to have great thoughts, rather than managing huge projects," he answered. "They have established organizations to handle such large projects under the university's direction. Perhaps you have heard of the Jet Propulsion Laboratory that is associated with Caltech or Lincoln Labs that serves that function for the Massachusetts Institute of Technology. Thousands of very competent scientists and engineers at such support organizations handle the development of instruments and all the associated management leaving the academics to focus on pure research and teaching."

The maid arrived with coffee and some pastries from the village bakery. Philip enjoyed one of his favorite "sow's ears" pastries while Estelle poured his coffee. He didn't mind answering her questions. Still he thought it might be better to change the subject.

Chapter Ten

"I hope you won't mind if I ask you about your life in this lovely place," he asked, somewhat timidly. "Not at all," she responded, settling herself on the sofa with her legs pulled up beneath her to project a familiarity intended to put Philip at ease. A family portrait hung behind her, showing Estelle and Bernard sitting on this same sofa with all three children assembled around them.

She noticed Philip looking at the family scene. "Yes, our children are my joy—they are both so kind and loving," she commented, turning her head to view them over her shoulder. "I also have my hunting and the supervision of the estate to keep me busy. And of course, I pray that Bernard will return to live here again soon."

"How are you managing all that?" he asked somewhat hesitantly but encouraged by her forthrightness.

"As best we can. In the past, you know," she began, "it was not uncommon for married couples like us to lead almost separate lives. Husbands would have their 'petites amies' in Paris and their wives felt free to also find young friends to keep them occupied.

"Back then such people often married partners chosen partly with an eye to keeping large properties or fortunes as intact as possible. So such arrangements seemed practical and even sensible. But it is less common now." "What has changed?" Philip asked. "It causes too much pain," she said quietly. Philip reflected on this simple avowal and decided not to pursue the issue.

"I am thinking," she said, returning to the topic of his research suddenly, "that if you ever need financial support for research that might be difficult to fund in your normal ways, you should consider involving a friend of ours who is very interested in astronomy and also very rich." He sensed that she had been inspired to mention this out of a sense of guilt over having complicated his life with her question. But he didn't want to remonstrate with her on that score any farther.

Philip thought it unlikely that he'd need to avail himself of her offer but it was true that even big science projects were increasingly being funded by private philanthropy. The two giant Keck telescopes that Caltech and the University of California built in Hawaii had already pioneered that trend in the 1980s. It was commonplace now. Many scientists were finding it easier to deal with a sympathetic patron rather than navigating the increasingly tortuous bureaucracy required to apply for Federal grants.

But it was time to thank her for Bernard's contact information and for her kind hospitality at lunch and this afternoon. She looked at him sincerely and touched his hand. "It has been a pleasure to converse with you," she said. "I am making an effort to meet interesting new people, and even to flirt a little," she laughed. "It is more normal with the French," she assured him, seeing that this last had taken him by surprise.

"When Bernard and I lived in New York for a few years, I missed the flirting," she explained. "We lived in an elegant apartment building in Manhattan and a few times it happened that I would meet in the elevator some man who had chatted with me at a cocktail party the night before and now he asked me whether I'd like to have a quickie! Can you imagine that? There seemed to be little feeling for civilized give and take. I mean the fun of just repartee and teasing playfulness between men and women."

Philip smiled and agreed, a bit nervously, that such uncivilized behavior was an unfortunate mark of our age. "Our time together has meant a lot to me also," he managed to get out in a slightly hoarse voice. Then he hurriedly excused himself, saying that he needed to prepare for departure after the other attendees returned from the excursion.

Soon afterwards the group arrived tired but abuzz with new impressions. They had visited the tomb of William the Conqueror in a medieval church at Caen and had enjoyed a gastronomic lunch that featured the local specialty, tripe soup. The trip concluded with a somber visit to Juno Beach.

The excursion had been a welcome break after a fractious morning and following Cecily's closing remarks the attendees were quite content to find their taxi rides and head back to their train stations and Paris. Everyone looked forward to blissful anonymity in their various hotels after a few stressful days of keeping up appearances.

Chapter Eleven

On the daytime flight home Philip was able to read over two papers that he had been sent to referee for the *Astrophysical Journal*. One was excellent and he enjoyed reporting on its virtues to the editor and recommending that it be published. He was less impressed by the other. Philip didn't enjoy writing negative reports, but this work was so sloppy that it upset him to have to waste time on crafting a rejection that was firm without being unkind.

His mood was improved by an interesting conversation with his neighbors on the plane. He was used to people asking him about his work when they found out that he was an astronomer. But Myrna, the woman in the next seat, did him one better.

She was a sea lion trainer who travelled the world with her boyfriend and a troupe of well-drilled performers. While she napped, her boyfriend confided to Philip that she had earned the highest math aptitude score in her graduating high school class. Myrna confirmed that later, smiling bashfully. "Everyone was putting pressure on me to study something that involved math. But I really wasn't interested. I just happened to be good at it." Myrna's story interested Philip.

He returned to Cambridge to a few research papers that needed finishing and some grad students who needed supervision. He didn't recall expecting as much hand holding as seemed to be the norm these days. But teaching his astronomy course for undergrad science majors would be fun. It was a full-year offering that he shared with a younger faculty member who was more familiar with galactic astronomy. That is, with everything that lies inside the Milky Way and other galaxies like it: stars, gas, dust, the sun, and the solar system.

A colleague had joked that, for cosmologists like Philip, everything that lay inside the so-called Local Group of galaxies, including the Andromeda nebula and a few others nearby, was just geology. For them, real astronomy was what happened on the vastly larger scales of the universe's overall structure involving countless galaxies out to unfathomably huge distances.

His joke was closer to the truth than many astronomers liked these days. When the huge energetic particle accelerator project called the Superconducting Super Collider collapsed in the 1990s, scores of high energy particle physicists left their field and re-tooled themselves as astrophysicists. Most went into cosmology, especially into using cosmological data to explore the nature of matter in different ways than they had done previously using high energy particle accelerators.

The transition was made easier because the phenomenology of cosmology is relatively sparse compared to galactic astronomy where every star's spectrum has a particular story to tell about its chemical make-up, rotation, heat transport mechanism, and movement in the galaxy. To do cosmology, you could learn most of what you needed to know about the phenomena of the extra galactic universe in a day of reading a few review papers. Armed with that information, you were ready to set math loose on the universe. The rest depended, it seemed, on how brilliant you were at deduction.

Many astronomers resented this invasion by physicists, spluttering about how they had no feel for the subject and absorbed too much of the National Science Foundation's astronomy budget pursuing fanciful model universes. Mostly, they felt threatened by their universities' appetite for hiring high-status foreign physicists, many recent emigrés from the Soviet Union, over the home-grown product.

Now that he had tenure, Philip didn't spend much time on such issues. He enjoyed teaching except when he felt under pressure to be attending to issues that tugged at his conscience. He had been away from home while Emily struggled with filling out all the forms and schedules required for school and college applications.

She had her own career as a biochemist at an industrial research lab and they tried to balance the workload averaged over time although there were stretches when one or the other had to pick up more of the burden. He knew that she was coming up on some proposal deadlines, so he planned to make time for family matters over the next few weeks.

Luckily, Philip was seldom required now to leave for a week or more on observing runs at distant observatories. For one, he was using data from space telescopes more often and also the postdoctoral fellows working with him were able to accompany graduate students for training on ground-based telescopes, mainly in Hawaii and Chile.

This made family life more relaxed. Emily asked a few perfunctory questions about his meeting in France, but she wasn't too concerned when he told her of his decision. She was even pleased that he might have more time to spend with the kids before they left the house for prep school and college. He hadn't seen the need to mention the Countess, although there was a chance that a colleague might tell Emily that Philip seemed to have been a mite distracted by conversation with their comely hostess.

Plenty of their faculty friends had experienced wobbles and re-directions in their careers; some had essentially run out of steam before they had even reached Philip's age. They had drifted off into the history of astronomy or into metaphysical ruminations on the Arrow of Time. The department couldn't shed them but they could be re-assigned to offices in the basement. Others were still plugging away on the same research topics in their eighties, generally to the consternation of younger colleagues.

The truth is that there are only so many well-posed problems in science at any given time despite claims to the contrary found in public pronouncements. The number of those problems suitable for big budget research projects favored by the funding agencies these days is even smaller. Having old timers hanging around and springing up at meetings with reminders that many of these problems had been solved decades ago was simply bad for business.

Luckily, the increasing use of social media rather than scientific journals to discuss and propagate research findings helped to marginalize the old timers who weren't nimble enough on Facebook, Reddit, and Twitter to keep up with information flow. After a few years of feeling left out they generally retired to focus on writing novels or gardening.

The half-hour drive in to Cambridge from their home in Belmont west of Boston gave Philip some time to think about how he'd open his first class of the spring semester. The students had spent since January break with his colleague who enjoyed observational astronomy. Philip was impressed that the younger man had organized some interesting lab exercises on a small telescope at the venerable Agassiz Observatory in Harvard, Massachusetts, about an hour off campus.

So Philip thought that he should connect to that experience and then, he thought, he'd talk about some of the underlying assumptions that make study of the universe possible. Like, first of all, the Uniformity Principle that all corners of the universe are made of the same elements and obey the same laws.

When the development of the spectroscope in the early 19th century made it possible to separate light from astronomical objects into its rainbow of colors, comparison of the patterns observed in the rainbows received from different stars and fuzzy nebulae with the patterns found in laboratory spectra of glowing materials or flames showed surprising commonalities.

After some clever analysis it was shown that all the spectral patterns could be traced to radiations from the same chemical elements and glowing bodies found in laboratories on Earth. That finding was an amazing and not at all obvious discovery. That discovery encouraged mankind to venture forth with confidence into exploration of the universe.

Yes, thought Philip, I'll spend the first day laying out the evidence behind the Uniformity Principle. It'll give me a chance to gauge their level of comprehension so I can plan how to pitch the next lessons.

Chapter Twelve

Philip's course was an advanced survey of astrophysics which drew about a dozen juniors and seniors, far fewer students than the Astronomy for Poets course that filled an auditorium. His students were mainly physical science majors who were planning to study those subjects in graduate school and aim for research careers.

Occasionally the course attracted some unlikely participants, like the football team lineman enrolled this year who "wanted to know what the universe was all about." Philip respected the judgement of such students who knew that they wouldn't ace this relatively challenging offering, but were looking beyond grade point averages in choosing their courses.

The first class went reasonably smoothly. Some time was, as always, taken up choosing the meeting times to fit everyone's spring term schedules. It seemed to Philip that this exercise required as much analytical aptitude as the course material.

He was impressed that the Teaching Assistant, a pale graduate student who took a seat inconspicuously in the back of the room, was able to cut the Gordian knot posed by the students' scheduling conflicts. He had been assigned this assistant by the department to help with papers and quizzes and it looked like he would be an asset.

After these preliminaries Philip launched into the discussion he had planned of the principles and assumptions, the axioms if you like, underlying study of the universe's birth and evolution. He outlined how the subject had evolved from antiquity to its present frontiers. Judging by the questions, it seemed that everyone was following these introductory ideas easily.

He described the Uniformity Principle and pointed out how its universal applicability might be questioned. For instance, the mass of galaxies seemed insufficient to prevent their rotating material from flying apart unless one hypothesized some form of unobserved dark matter. The alternative would seem to be violating Uniformity by adjustment of one of the Laws of Nature, the inverse square law of gravitational attraction.

He also drew attention to weaknesses in our understanding of the very nature of reality suggested by incongruities in quantum mechanics. The century-old controversy between Albert Einstein and Niels Bohr on the interpretation of quantum mechanics remained alive. There was still no good answer within the range of our experience, as to how phenomena seemed to be able to interact instantaneously over cosmic distances.

Some of this was getting in a bit too deep, he could tell by the class's body language and wandering gazes. But he still wanted to impress them with the amazing fact that even very simple deterministic systems could exhibit chaotic behavior.

"Henri Poincaré had already shown over a century ago," he began with forceful enthusiasm, "that systems where the effect feeds back on the cause, i.e., non–linear systems, could exhibit unpredictable behavior. But it wasn't until the 1960s that a MIT meteorologist, Edward Lorenz, brought this finding into the limelight."

He could see that mentioning the place down the street brought the flicker of recognition he was hoping for. "Using the equations of a leaky waterwheel as an illustration," he continued, "Lorenz and others showed how such a simple and deterministic system could exhibit weird behavior. The waterwheel might suddenly and unpredictably change its direction of spin, for example." A low hum expressed that the little boffins were impressed.

"More generally, such childishly simple systems, whose physics was completely described by Newtonian mechanics, could develop chaotic behavior," he forged ahead. "The best known application these days is the observation that the beating of a butterfly's wings in Brazil can determine the behavior of a tornado in Texas." Smiles flickered; the students were relieved that he was touching back to something that they had heard of before.

Dismissing the class, he hoped that he hadn't frightened any of them away. To further calm their fears he assigned a relatively easy piece of homework—to write a short essay on some other assumptions that they thought might influence our ability to understand the universe's birth and evolution. They looked like a smart bunch; he was curious to see what they would come up with.

Chapter Thirteen

Anna Grad had impressed Philip with her grades in the introductory physics and math courses in her junior year and towards the end of the spring term she had come to his office to ask whether he might consider employing her as a research assistant over the summer. Philip was well aware of her quickness. More often than not she had arrived at a point he was laboriously constructing in class before he had. When this happened he wondered whether she was as precocious as she seemed or whether he was just getting slower.

It reminded him of a mathematics colleague's comment that there was a price to pay for quickness of mind. "Whatever I'm going to contribute to solving a problem," he once said, "will come out in the first twenty minutes. I'm quick but not deep. That's why I seek out collaborators whose minds work more slowly but more profoundly." Philip comforted himself with the thought that he tended to be more profound.

As the summer wore on Philip recognized that he looked forward to their sessions together examining data and discussing its analysis. She seemed to like him and he even had to admit to himself that he found her attractive. Anna was not a classic beauty but she was quite tall, brunette and had a nice figure. Her green, somewhat slanted eyes gave her an exotic look. He admired her graceful, long fingers as they leafed through printouts together. He thought about her probably more than was good for a happily married older man.

Sometimes when she was trying especially hard to make a point Anna had a habit of coming over and sitting sidewise on the edge of his desk, dangling her long legs near his. This made Philip uncomfortable and at first he used to roll his chair back to get some space, but gradually he became more accustomed to her behavior, chalking it up to her Russian background.

Anna's parents were Russian Jews who had immigrated first to Israel and then to the US when she was in elementary school. They were both biologists working in academia in the New York area. Anna was largely Americanized but occasional slips in the order of words or incorrect use of articles in her sentences betrayed her Russian heritage.

Philip was well aware of the perils of consorting with young women in the charged atmosphere that had been generated over the past few years. So he always met with Anna during hours when others were around in the department. He also made a point of keeping the door to his office wide open. When Anna asked about his personal life he answered briefly but steered the conversation back to research matters.

None of this deterred her. Despite her intellectual sophistication, socially she was surprisingly childish in a way that Philip found appealing. She mentioned in passing, giggling, that she was teaching herself Classical Greek so she could read Sappho's love poems in the original, but then confessed that she had few American friends of her age because they found her naïve.

The arrangement went well and continued into Anna's senior year. She asked whether he'd be her senior thesis advisor and he said that he'd be happy to do that. They even wrote a short paper together. That was fairly standard these days, to give undergrads a leg up, so to speak, before graduate school.

Anna did once ask whether Philip might take her on an observing run to Chile or Hawaii; an astronomy major she knew had done that with his thesis advisor. But Philip wisely thought better of that. He didn't want to tempt fate and he'd need to tell Emily about Anna after over a year of keeping her to himself.

Around Christmas Anna told Philip that she'd heard about some one-year Congressional Internships that NASA was offering for graduating seniors in science interested in making a difference in Washington. She thought that she might give this a try instead of going straight to graduate school and wondered whether he might write her a reference.

He was happy to do that too although it saddened him to think that she would be leaving soon.

A couple of months later Anna rushed into his office afterhours waving a letter and exclaiming, "I've got it! I got the internship!" Breaking the decorum that they had both adhered to she flung her arms around his neck and kissed him on both cheeks.

Philip had no time to react but managed to look past her as soon as he recovered, to assure himself that no one passing in the hall had witnessed this outburst of youthful enthusiasm. On the surface he smiled nervously but inside he remembered the statement for which an eminent English biologist had recently been pilloried: "The trouble with having young women in one's laboratory is that one tends to fall in love with them."

Chapter Fourteen

Philip met with the Astronomy Department chair and the Director of the Harvard-Smithsonian Center for Astrophysics (CFA) a few days after returning from France to discuss his withdrawal as chairman of the new Colossal Optical Space Telescope (COST) project. As he expected, they didn't make much of a fuss; both expressed confidence that the time freed up might be better employed in taking on some additional administrative tasks in the Department.

They seemed mainly interested in his recommendations for someone else at the Center who might take over. NASA still wanted the new leadership to be from Harvard, and the CFA wanted to keep an inside track on the study phase funding. Philip said that he'd give it some thought and get back to them soon.

It was a pleasant early April day, so after the meeting Philip walked down Mt. Auburn Street and over to the Science Center to meet with his class. It was one of those times, more rare as he aged, when no part of his body hurt and his step was jaunty. Plenty of students were enjoying the sunshine on the plaza outside the Center.

He tried not to stare at the co-eds sporting the favorite new attire: skin-tight black leggings. As the father of a teen-aged girl, he sympathized with efforts to curb sexual harassment, so it seemed a strange time for women of all shapes and sizes to favor a form of dress that on many wearers seemed to him so brazenly suggestive.

On this confusing thought he entered the classroom where his students were already waiting. "Eager to learn, are you?" he asked the class, grinning. They grinned back. As far as he could count, no one had dropped the course yet. It was going to be a good term.

Philip wanted to introduce them to one more example of the surprises that cosmology had in store before he moved into the more formal parts of the course. “A really interesting finding of the last few decades that I want to tell you about,” he started, “is the so-called fine tuning of the universe. Perhaps you’ve heard that term. So what does it mean?” he asked rhetorically.

“Well, if you take ratios of certain pairs of fundamental constants of nature, like the Coulomb force between two charged particles and their gravitational attraction, you get a set of dimensionless numbers that have been measured to reasonably high accuracy. It turns out that the universe seems to be amazingly sensitive to small changes in these numbers.

“In the case of the ratio of the Coulomb repulsion between protons to the so-called strong force that holds atomic nuclei together, if the value were different by about 10% from the measured value, life as we know it on Earth could never have developed. The reason is that this ratio determines how well hydrogen can transmute to helium.

“If the strong force were weaker than it is, helium couldn’t form and the universe would consist of only hydrogen. Since the stars derive their energy from burning of hydrogen to helium and heavier elements, there would be no energy source to support life.

“If the ratio were higher, hydrogen would have formed helium *too* efficiently in the early universe and no hydrogen would remain to fuel stars over the long time-scales required to produce life. Similar consequences are found for small deviations of five other dimensionless ratios. So it looks like the universe is very finely tuned. Strange, isn’t?”

The class was silent. Philip could see that this was something new and that they were impressed. “This fine tuning seems to have interesting implications for our understanding of the universe,” he went on. “Two main explanations have been put forward. One that you’ve probably heard about is the concept of multiple universes.

“Here the idea is that, if you have a sufficient number of parallel universes each with different values of these ratios, then by chance one will occur in which the ratios have the values needed to produce our universe. It has been suggested that 10^{37} or more universes might exist!”

"The other main interpretation is that the ratios have been fine-tuned by some entity to the measured values." He stopped to let that sink in. "That is the Divine Creator explanation," he pronounced with some grandiloquence, watching for the effect on the class.

Philip could see that they weren't sure how to react to this rather profound revelation. He asked if there were any questions. One student requested some sources to read more about all of this. Philip suggested Martin Rees' excellent monograph "Only Six Numbers" and told them that he'd assign a forthcoming piece of homework on that book so they should all read it over. Then Philip pulled out his notes and plunged into the derivation of Einstein–de Sitter universes. The class knew that the fun was over and the cosmology portion of the course had begun in earnest.

That evening Philip started reading over the short essays he had assigned on the Uniformity Principle. There was a wide range in the maturity of the work. The author of one of the weaker essays wasn't grasping what the Principle was about. He pointed out that many stars had unusual chemical composition so he didn't see how the Principle had any generality. Philip was not sympathetic because the differences in stars' compositions had been explained to the class by his colleague during the previous term.

The last essay that Philip had the energy to read before turning in for the night was by Anna Grad. She asked whether any doubts had been expressed as to the ability of the human brain to understand the universe. Moreover, she continued, was there any evidence that the brain had evolved in such a way as to preclude it being able to understand that universe?

"Interesting questions," said Philip to himself as he slipped under the covers.

Chapter Fifteen

Emily's family had a summer place in Maine on Mount Desert Island and they spent a week or two there every summer catching up with relatives and with Emily's childhood friends from Philadelphia. Philip looked forward to these stays although it was a long drive.

He found it interesting that various places along the Maine coast seemed to be frequented by people from specific US cities. The rusticators, as the summer people liked to call themselves, in Northeast and Seal Harbors on Mt. Desert hailed mainly from Philadelphia and New York City. The pretty island of North Haven a bit down (actually "up" in a sailor's way of thinking) the coast attracted the Cabots and others from the Boston area. Biddeford Pool was chock full of jolly Southerners from Richmond, Virginia.

The weather was unpredictable; some years they never saw the sun except in brief snatches. Just fog and drizzle. Last year, Philip remembered, had been glorious. It even got hot enough to plunge into the sea although only mad dogs and Englishmen, as the saying goes, tried to impress by swimming out as far as the lobster buoys a hundred yards or so offshore.

The family had a little green sailboat, named *Goblin*. She was a venerable wooden, gaff rigged, 16-footer that Philip liked to sail around the bay and out to Little Cranberry Island. Her century-old design by the famous marine architect Nathaniel Herreshoff was still eye-catching. Philip had maintained a wooden sailboat as a teenager and knew all about sandpaper and varnish but he was happy that *Goblin* was cared for professionally. The ultimate decadence, he thought, was that she was even launched and hauled by a local yard.

When the kids had been younger, Philip had enjoyed teaching them how to sail but in recent years they had been enrolled in the junior sailing program at the little grey shingle yacht club perched on a rock sea wall over Seal Harbor. They preferred sailing with other young people and Philip understood that perfectly. He and Emily were just pleased that they were outdoors and not staring into a computer screen.

Philip sometimes tried to persuade Emily to join him afternoons after the sea breeze came up, but she preferred to just visit friends or curl up with a book. Once in a while Philip managed to find someone to join him, but only on perfect days. Just a few of the truly dedicated cared to sail around in the grey mist unless a race required it.

This year they had the house for a couple of weeks in early July. Philip was happy because it was still relatively quiet then—the smart set and their yachts didn't crowd in until later in July and early August. Emily and he had little in common with the financiers and captains of industry who jammed the town briefly and turned its casual social life into a frenzy. Many of them were interesting people but they focused their energy on others like themselves with the influence and money to further their business and philanthropic causes. They had limited time for a professor of astronomy, even one from Harvard.

Philip always brought some summer reading along. This year he had chosen a couple of books on the structure and evolution of the human brain. He knew next to nothing about biology, never having studied the subject even in high school. But he was bothered that he had no answer to the question raised by Anna: "How much confidence could we have that the human brain is even able to comprehend the birth and evolution of the universe?" she had asked. He had resolved to investigate over the summer whether neurology, archaeology, and anthropology had anything to offer on that question.

Over the years Philip had noticed a pattern in how his colleagues approached their research. Some had been acknowledged wizards since elementary school; in high school they were of the small group that could solve deductions in geometry that no one else could figure out. They seemed to understand concepts like logarithms before the teacher even had a chance to explain them.

These were the individuals whom college professors spotted right off and talked about at faculty meetings. "We need to get him (or her) back here after grad school before MIT gets him," they'd say. Some of these wizards, like the Englishman Martin Rees, were solving problems that puzzled their older colleagues while they were still undergrads. Some, but not all, made major discoveries later in life.

These individuals were adept at deductive thought, as it is known. Presented with a few axioms they could produce solutions that few others saw. But as it happens, relatively little of science is amenable to solution by such deductive brilliance. Richard Feynman, perhaps the most famous of such genii in Philip's time, decided after getting a Nobel prize in physics to try his hand at biology. Soon he decided that was a mistake. "Biology is not ready yet for a mind like mine," he reputedly commented.

Feynman was alluding to the fact that biology is a relatively messy subject that requires a mind able to absorb a welter of disparate facts and synthesize them. This process of proceeding from the general to the particular is called inductive reasoning. It is, in a sense, the opposite of deductive reasoning, as the Greeks were already aware. Astronomy, Philip observed, lies intermediate between the purely deductive discipline of mathematics and a highly inductive subject like botany. Occasionally it can be useful to have strong deductive skills in astronomy.

For instance, the celebrated demonstration from basic fluid mechanics that expansion of the sun's atmosphere would produce a supersonic solar wind was widely disbelieved until the first space probes detected that wind several years later. Now this model, derived by Professor Eugene Parker of the University of Chicago, is the basis of solar terrestrial physics. Yet, Parker had once freely admitted to Philip that his powers of inductive reasoning were limited.

Philip saw a possibly interesting application of this duality in how scientists' minds seemed to be arranged, to the issue of gender differences in math and science aptitude. But after watching how Harvard's President Larry Summers was hounded out of office after daring to suggest that such differences might favor men, he decided as did many others that it was best to stay clear of the topic.

Still, it seemed to him that the successful female scientists he knew used mainly inductive skills to arrive at their results. The feats of deductive cleverness that produced so many of the paradigm-changing theoretical advances in astrophysics were mainly contributed by men, even though Harvard already produced more female than male astronomy Ph.D.'s in the 1920s.

The women's contributions, like Cecilia Payne-Gaposchkin's amazing discovery in the 1930s that the sun was made mainly of hydrogen, could be as important as any that men had made. But they relied more on inductive skill, so they seemed to be less valued amongst academics.

This fascination with deductive over inductive cleverness was not a predictable trait of the human mind, Philip thought. What Darwinian advantage did it confer, after all? He wondered whether it and the gender difference in deductive and inductive aptitudes might provide clues to our mind's ability to grasp the nature of the universe.

Chapter Sixteen

That summer was of the foggy and drizzly variety so Philip had plenty of time to read. He soon learned that, for all the interesting advances in human evolution over the past few decades, the development of the brain's capabilities remained poorly known.

The basic problem was easy to understand. The hominid fossil remnants gave evidence on evolution of the human frame and even muscular structure. But except for some evidence on brain volume, there was nothing to go on to deduce its capability for thought. That could only be studied indirectly through investigations of tool making ability, social structures, and so on.

Recent advances in so-called functional magneto resonance imaging, or fMRI, did seem to be showing that the male and female brains were wired differently. So there was hope for untangling such gender difference issues in the future, but much work remained before one could hope to assess the brain's ability to grasp cosmological questions and determine what role evolution might have played in forming that capability.

Disappointed in his hope that he might improve his response to Anna's question, Philip spent more time at Seal Harbor that summer with Emily and the children. Becky, their eldest, was going to Wellesley College in the fall. She was capable and socially adept. He and Emily expected that she would continue as an achiever and leader much as she had been in high school.

Their son William was more challenging. Philip and Emily knew that he was at least as bright as Becky but he had more difficulty adjusting socially. He avoided group activities and participated reluctantly even in individual sports like tennis or skiing. He wasn't prospering in the local public middle school so they had decided to try a boarding school. They thought it might help to give him some separation from his family.

Both of the children were, as today's norm went, well educated. But Philip had doubts. Neither they nor any of their friends seemed to have much of a vocabulary. They read for school but little otherwise, compared to Philip's recollections of his reading habits at that age.

They also lacked much idea of geography, or even where they were themselves, at any given time. Maybe GPS navigation is partly to blame, but the way we educate them is at fault too, Philip thought. When Philip was William's age, he remembered, he knew that the capital of Honduras was Tegucigalpa. He doubted that William knew the capital of Canada.

Their background in history seemed to be limited mainly to the Holocaust and the Civil Rights movement. Neither of them (nor, he thought, most of his Harvard students) could have told him in which centuries Napoleon Bonaparte or William Shakespeare had lived. This lack of any real content in their education was traceable to the idea cherished by educators since the 1960s, that there was no point in teaching facts—it was better to teach kids how to find facts.

The problem that Philip saw with this cheerful piece of misguidedness was that creativity relies on knowing facts and being able to link them. The mind needs this matrix of seemingly unconnected trivia to perform the connections that move mankind forward. Going looking for them afterwards doesn't help much. Sometimes the connections are made when we are awake and sometimes when we are sleeping. But without the underlying matrix, creativity lacks the material it needs to function.

Emily was less invested in such doubts; she was just grateful that the kids weren't on drugs and that she and Philip weren't being called by the school or by the police. Many of their friends in Belmont, she pointed out, weren't so lucky.

They knew that they had been fortunate in having at least this satisfaction, but they also felt that keeping the kids firmly grounded had played a role. William was the only boy in their neighborhood who mowed his family's lawn. Everyone else had landscapers do it while their children played computer games. Betsy helped clean the house and both kids took care of clean up after dinner. Little things like that reminded them that they had responsibilities and made them appreciate their free time.

Philip felt that, with the kids leaving the fold, it was important this summer to create good memories of family life. The results were mixed. One evening when the weather looked fine he borrowed the sailing club's outboard skiff and took the family over to dinner at a popular seaside restaurant across the Sound on Little Cranberry Island. The food was good and the kids were cheerful, but when they emerged to return it was getting dark, the fog had set in and visibility reached barely beyond the end of the dock.

No one had thought to bring a cell phone so they couldn't navigate home by GPS. In fact there wasn't even a compass on the boat. Philip felt that he had a pretty good sense of direction so he thought he could get them home by dead reckoning. He headed in the direction he thought they had come from but after the expected time had elapsed there was no sign of anything in the fog, not even distant lights. To make matters worse, Emily checked in the stern and found that there wasn't much gas left.

Philip's mind raced over what he remembered Polynesian mariners used to navigate. Migrating birds wouldn't be helpful tonight he thought, but wind and wave direction might give a clue. He remembered that, on the way over, the waves had been coming at them from the port bow. The tide hadn't turned so if he oriented the boat now to take them on the starboard stern they should be headed back the way they came. He hoped that they hadn't gone too far off course.

They were very relieved when, after another ten minutes the fog cleared and they found themselves not too far down the coast from their yacht club. He knew that he'd hear about this little adventure from family and friends for the rest of his life. But it convinced him that higher powers of reasoning did provide a Darwinian advantage. It wasn't clear whether he had brought them home mainly by deduction or induction but he was content to leave that open. Whichever it was, if they had run out of gas at night with the tide sweeping out to sea, the story might not have had a happy ending.

Chapter Seventeen

Back home in Belmont the Frobishers didn't attend church regularly. But although neither Philip nor Emily were religious they recognized that, even now when the government had assumed most benevolent functions like running hospitals, the churches still had great value to the elderly and less fortunate members of their community. So they made an effort to show up at very least for Christmas and Easter and on the birthdays of deceased loved ones.

Here in Maine they attended more often because it was also an opportunity to catch up with family and friends. Philip was sorry to see fewer younger members of the families, like Emily's, who had been summering on MDI for generations, attending the services. Their presence had always attracted a flock of the nouveau riche who in turn helped to fill the pews with those of more modest social stature. That had probably been a feature of church going since the Middle Ages, he supposed.

Philip liked the stained glass windows in the 19th-century chapel and did his best to make out the harmonious words inscribed on them. He wondered how those particular calm phrases had been chosen long ago. Occasionally a sermon struck a chord in him. Reverend Larry, as he liked to be called, was a spare figure of about Philip's age, down from New York for the summer.

One sermon in particular caught his attention; it dealt with the central importance of faith in Christianity. The example chosen was the difficulty of believing in Christ's resurrection and whether it could be understood within the scope of modern physiology. Faith was needed, the minister suggested, to close the gap. Reverend Larry concluded that sermon by noting what a fine day it was to get out and enjoy Nature. It looked like an especially nice breeze for a sail, he added with a smile. Philip took this cue to invite him out for a spin in *Goblin* after lunch and the Reverend accepted.

When he arrived at the yacht club, Larry was already waiting. Philip was relieved that he was dressed like he had been around boats. They pushed the little dinghy off the dock and Philip got in first, to row. Unfortunately, the minister was substantial enough to tip the bow upwards when he slid himself off the dock into the stern, soaking his shorts. Philip apologized that he should have thought to move forward but Larry assured him that he didn't mind. With anyone else Philip would have insisted that they go back for dry clothes but somehow he felt that a man of God knew best.

It turned out that Larry even knew how a gaff mainsail was rigged so they had *Goblin* ready in short order. They cast off with dispatch and soon had her gurgling along cheerfully on a beam reach up the shore. Larry thanked Philip for the offer to take the helm, saying that he'd rather watch for a while at least.

Larry told Philip that he'd heard that he was an astronomer and that he had always been fascinated by advances in understanding the cosmos. "Especially doctors seem to be closet astronomy buffs," Philip confided. "Many tell me that they would have rather gone into astronomy if they hadn't flunked calculus." Larry laughed, saying that he had passed calculus but still felt that his calling was elsewhere.

Goblin dipped and rose over the long swells coming in from the open ocean. It was one of those precious days when the breeze and tide conspired to push her along, rushing down each passing sea. The water sparkled in the bright sunshine. It was a good time to discuss the state of the universe.

When Larry asked what was new in cosmology Philip was happy to oblige with the finding that he had described to his class that spring, of the strange fine tuning of our universe. "That is certainly remarkable," agreed Larry, after listening attentively. "I suppose you expect that I'd opt for the Divine Creation explanation over the parallel universes."

"There is a third interpretation that cosmologists are reluctant to discuss," Philip added, letting out the mainsail a bit as the wind freshened.

"Our minds may not be able to comprehend the make-up of the cosmos. We may just be flailing around, unable to get any traction on a problem that simply lies outside our intellectual capability.

"Of course it's very hard for scientists to admit that possibility, given the great strides man has made in science over the past few centuries. But there seem to be limits to our knowledge closer to everyday experience that have been revealed recently," he continued.

He told Larry about another one of the discoveries that he'd discussed with his class, that the non-linearity of most systems found in Nature, the simple feature that the effect only rarely failed to act back on the cause, made those systems prone to unpredictable, chaotic behavior. "This chaotic behavior limits what we can learn about those systems, like the path of a tennis ball, or climate, regardless of how carefully we measure their properties. It's a chastening message that hasn't been fully absorbed yet in the science community," he said, meanwhile peering out to sea where the sky was darkening. "Let's go about and head back," he suggested, "the weather seems to be changing."

Larry handled the jib deftly without catching the sheets on the mast cleat and they turned *Goblin* around for the trip back. Happily the wind was farther aft now so they could let the sails out and scud along towards home without misadventure. Emily liked to say that sailing was boring except when it was terrifying. Philip only smiled forbearingly when she came out with statements like that. They did share lots of good times in other outdoor activities like tennis where their limited skills offered equal challenge.

On the way back, Philip asked Larry to tell him about the topic of his sermon—faith and its importance in religion. He said that, to him, that term had always sounded hollow. Yet, about forty percent of Americans preferred to believe in Creation over advice that the beginnings of the universe and of humankind were accessible only through scientific research. Philip felt that, in the face of such a statistic, understanding what underlies faith might provide another clue to our mind's ability to comprehend the universe.

They were approaching the yacht club so Larry had little time to elaborate, but gazing at *Goblin*'s wake bubbling astern, he told Philip: "To me, faith is a matter of trust. You trust in something like the resurrection if you believe that your faith in it will promote the virtues that Christians wish to represent. I don't know how satisfying that is to you, Philip, it's just my personal take on faith."

Philip nodded. He wasn't convinced that Larry's explanation covered the entirety of human faith. But it did sort of agree with his own belief that people's trust in science, faith in it if you will, stemmed more from their trust in scientists than from any attempts to understand complex technical arguments.

He felt that a scientist who managed against all odds to have his life together was better able to convince his or her community of global warming or Darwinian evolution, than the most passionate activist. So, he thought to himself, the connection that seemed to exist between faith and trust might help promote a reconciliation between outlooks that were dividing the nation and the Western world. Pondering this portentous thought in silence, he rowed the two of them back to the dock.

Chapter Eighteen

The Harvard College Observatory occupies a pleasant hilltop within easy walking distance of Harvard Square. It borders on a Cambridge neighborhood of fine homes and it has always seemed a good policy to maintain friendly relations with their owners.

In the fall after their return from Maine the Frobishers were invited to a little dinner for some of the more well-to-do members of that community. Such events were held from time to time to introduce potential donors to exciting scientific developments and to star faculty and students whose work might benefit from direct financial support.

Philip and Emily found themselves seated with the Observatory Director, Nancy Stein, another faculty couple, a graduate student and two of the Cambridge neighbors. After introductions around the table, Nancy mentioned that she had just returned from some summer conferences including one held in Prague.

She gave a glowing description of the city and of how it had changed since she'd been there over twenty years earlier. Several others chimed in with their experiences in Prague and in other Czech towns like Česky Krumlov in the foothills near the Austrian border. One of the neighbors seated across from Philip, whose foreign sounding name Philip had not caught, said that he had been born in Prague and maintained a residence there.

"It is an interesting country that we hear little about because, like our neighbor Canada, it causes no one any trouble," he commented wryly. "Yes," added Nancy, "they seem so accomplished technically and in sports, but the newspapers focus more on their more troublesome neighbors, like Hungary."

During the dinner the company found out a bit more about their fellow guest, whose name turned out to be Ivan Vesely. He was an elderly man of average height and still powerful physique. His black hair was combed straight back in a style commonly seen in 1930s movie stars.

His family had immigrated to Canada in 1948 after the Communists had taken over the Czech government. His father's company had been nationalized and he'd been informed that there was no future for capitalists like him. Luckily their family had been able to emigrate legally to England and, a few months later, to Canada. Ivan had studied chemical engineering at McGill University and had obtained a Ph.D. at MIT. He stayed in the US and had done very well for himself as an entrepreneur in that field.

The other invited neighbor asked the female graduate student about her experiences as a woman in science. The knowing smile that she cast around the table signaled that she'd often been asked that question and she replied perfunctorily that she was enjoying her work.

"The majority of the other students in the Harvard Astronomy Department are actually women," she added. "In fact," she continued, "I've felt awkward sometimes because it seemed to me that female science students starting in high school were given opportunities not open to boys in my classes who were often more talented than I was." This answer surprised her questioner and generated some awkward silence.

Blessedly it was time to listen to a few short talks by the featured speakers, and afterwards Philip and Emily moved over to sit with Ivan over dessert and coffee. She found him charming and Philip was interested in his experiences combining research and commercialization in a small business.

"A couple of smart engineers joined me and combining research with development of commercial products was part of our vision from the beginning. We were even able to attract funding from the federal Small Business Innovation Research Program to produce cutting edge products for domestic and export markets.

"That sort of arrangement would have been impossible anywhere else in the world," he continued. "Even in Canada, there isn't the pragmatism and flexibility that is a great strength of the US. The federal agency program directors were allowed to fund research that received strong reviews, wherever it was carried out," he continued.

Philip commented that there were times when he dreamed of commercializing some of the techniques and software that he'd developed for multiplexing detectors and analyzing huge astronomical data bases. Ivan laughed. "You should be grateful that you have tenure at Harvard; running a high-tech firm like ours is not for everyone."

"How was it that you have returned to Prague?" asked Philip.

"I am looking to sell the firm; we have been getting into mainly biomedical imaging whose applications I scarcely understand, and I began looking for potential investments elsewhere," Ivan explained.

"Beginning after the so-called Velvet Revolution in Prague I saw opportunities in real estate and other kinds of placements. I started by buying some apartment buildings, finding alternative lodging outside the city center for the tenants, gutting and re-doing the once lovely buildings, and selling them as condominiums for many times my outlay."

Emily inquired whether he had a family here in Cambridge and Ivan told her that his wife was visiting relatives in Prague. "We spend quite a bit of time over there," he added. "I still participate in investments, mainly trying to keep businesses in Czech hands. So much has been snapped up by foreigners."

"How did you come by your interest in astronomy?" asked Philip, partly out of genuine interest and also playing Harvard Development officer.

"I've always liked fine optical instruments; I collect antique telescopes and enjoy viewing objects in the night sky. But recently I've become more interested in cosmology," he answered to Philip's delight.

It was time to leave but Philip and Ivan exchanged contacts. Emily was pleased to learn that Ivan and his wife, Maika, played tennis. "We should get together for a game when Maika gets back," she ventured. "Absolutely!" answered Ivan. "We belong to a convenient little club hidden away off Mt. Auburn St. We'll give you a call when she returns so we can play before they close the clay courts for the winter."

"I'm glad that I came tonight and had a chance to meet him," said Emily as they walked to their car through piles of fallen leaves. Philip agreed. He saw interesting potential ahead.

Chapter Nineteen

Nancy Stein shared Philip's interest in getting Ivan involved at the Center for Astrophysics. The day after the dinner she called Amanda Bigelow, the head of the Harvard Development office, to tell her about an intriguing potential major donor. Amanda brought up Ivan Vesely on her donor software while they were on the phone and searched for his business and philanthropic involvements.

Surprisingly, her most current database showed no record of any board memberships, either of a business or philanthropic nature. Nor were any substantial gifts to non-profit institutions listed. Furthermore, he did not seem to belong to any of the prestigious Boston area clubs, like the Union Club, the Brookline Country Club, or the Eastern Yacht Club in Marblehead. This was unusual for an individual of his estimated net worth. He did belong to the quietly exclusive Cambridge Skating Club.

After a moment of silence to collect her thoughts, Amanda asked Nancy to give her a couple of days to check on Ivan's background in more detail. In this day and age you couldn't be too careful, especially when dealing with prospects of foreign origin. But she agreed that, if he looked clean, the CFA might want to invite him to join their Board of Overseers. With his net worth, proximity to the Observatory, and technical background, he looked almost too good to be true.

A few weeks later, Ivan called Philip to say that Maika was back from Europe and they arranged a date to play some mixed doubles the next weekend. The weather had been wet recently but that Sunday afternoon the autumn sun shone brightly through the trees overlooking the Charles River as Philip and Emily walked from their car. The Cambridge Skating

Club was a charming little spot hidden unobtrusively behind a stockade fence among pleasant residences.

The décor was very New England old shoe; well-worn rugs and furniture and members who looked chosen to match; a cheery group, mainly sprightly elderly ladies having tea; and a few tweedy looking gentlemen with rosy cheeks. Philip remarked to Ivan that the scene looked like something out of a BBC set for an Agatha Christie mystery. Ivan laughed, "Yes, they are all fine fellows, but to quote an Etonian friend of mine referring to his club in London, you wouldn't necessarily want to find yourself alone in the shower with some of them."

Ivan introduced Maika to the Frobishers. She was an attractive, sixty-ish but sporty looking blonde with a warm smile and a slight European accent that Philip assumed must be Czech. They chatted only briefly about her trip to Prague because their court was waiting.

Emily was relieved that the courts were oriented so the afternoon sun wasn't shining in her eyes. She'd never mastered playing tennis with sunglasses and wanted to do well and make a good impression. She noticed with apprehension that Maika walked onto the court with the loose-limbed gait of a natural athlete. She hoped that she and Philip would be able to hold their own.

It was clear during warm up that the Vesely's were more proficient and were placing the ball to make Philip and Emily look good. After the first somewhat lop-sided set, Emily suggested that they split up. In the second she and Ivan eked out a close win, and everyone was happy to have hit at least one brilliant shot. Their time was up, so they joined others in the club house for afternoon tea and some warm blueberry scones. Philip complimented them on their game and Maika admitted that she had played competitively as a youngster in Czechoslovakia.

"I learned my tennis from Mr. Koželuh in Montreal," said Ivan. "He was a wiry old Czech guy burnt nut brown by the sun. He'd been the world professional tennis champion for several years in the late 1920s and early '30s, had emigrated like us in 1948, and was giving lessons on public courts in Montreal when I was a kid. He didn't believe in building self-esteem the way US coaches do. He watched you play for a few minutes and if he detected a lack of talent he suggested that you take up volleyball instead!"

Emily asked why Czechs were so good at tennis. "It is interesting," agreed Ivan, "that per capita of population Czech-born players have won far more majors than any other nation. My theory is that it is a consequence of the unique Czech system of tournaments. There are almost two thousand clubs, each fielding a co-ed travelling team.

"The talented young men and women on these teams often go on to marry and have tennis–talented children," he went on. "Look at how many Czech-born greats had highly ranked parents: Lendl, Hingis, Sukova—all were children of highly ranked Czech players. In the US, collegiate men and women players have separate schedules and scarcely ever meet."

"I'd always thought it might be their desire to travel abroad when few were allowed out under the Communist regime," Philip commented.

"Sure, but other Iron Curtain players shared that motivation and didn't produce similar concentrations of talent in Poland or Hungary," Ivan pointed out.

Emily and Maika's conversation moved into discussion of their children, and of their schools. William was applying to a boarding school in New Hampshire and Emily was interested in the Vesely sons' experience there.

The women's pre-occupation with educational issues gave Philip an opening to sound out Ivan a bit more on his possible interest in some of the research at the Observatory. He was aware that getting potential donors "involved" was the first step taken by the university development office on the way to asking them for a donation. He felt a bit transparent going down that path but the size of Harvard's huge endowment recommended the validity of that basic approach.

Ivan recognized where the conversation was going; he had heard approaches like that many times before. He said right off that, quite honestly, he didn't play by the usual rules. He wasn't interested in joining a Board or giving to the Observatory's general budget. "To tell you the truth," he told Philip, "I've found that I'm not temperamentally well suited to functioning on Boards. If I'm not in sole charge I'm not happy and don't contribute well.

"I'm only interested if we can identify a truly unique project that excites me," he went on, letting the words hang in the air. "I can understand that," Philip acknowledged. "To tell you the truth, I've also been looking for something novel to rekindle the enthusiasm for research that I had twenty years ago."

As they rose to put on their coats and leave, Emily noticed an old black and white photo hanging on the wall showing a jolly scene of skaters on a rink covering the area of the tennis courts. "They still flood the courts and it doesn't seem to harm them," Maika told her. "We'll have to come out a give it a whirl in the winter, if you enjoy skating," she added. The prospect of another common activity with Ivan increased Philip's confidence in a fruitful relationship.

Chapter Twenty

When Cecily checked her email on the flight home from France, she found a note from her boss saying that she should read the attached letter. It was from a Bible-belt Congressman addressed to the Director of NASA informing him that he and several colleagues had heard about NASA's plans to request a large boost in the agency's budget to fund the Colossal Optical Space Telescope mission. It went on to say that the undersigned intended to vigorously oppose the unprecedentedly large request for a mission whose main aim was to pursue a "narrow and unproven path" towards understanding of Creation and the origin of mankind. The letter then quoted several passages from the Bible to demonstrate the wrong-headedness of NASA's approach.

At first Cecily was inclined to consider this threat as a joke that could not possibly be taken seriously in the United States in the 21st century. But then she remembered the delays in permission to build the large new telescopes on Mauna Kea and Haleakala. Those delays had dragged on for years because of protests and legal action by a few hundred Hawaiian activists taking exception to new construction on these sacred summits.

If a small group of native Hawaiians could cause such havoc to plans for two giant telescopes it would be naïve, she thought, to underestimate the power of these Congressmen. It was widely known that such initiatives to fight research that touches on Creation, evolution or even global warming had the backing of roughly one hundred forty million fundamentalist Christian Americans. She conveyed this in an email to her boss and received his approval to consult with her advisors.

He also recommended that she talk to a few Congressmen in the Bible belt whose districts included some of the NASA Centers like

Marshall Space Flight Center in Huntsville, Alabama and Johnson Space Center near Houston, Texas. These centers might benefit from this mission. He'd also like her to put the issue to some contacts at major aerospace firms like Martin Marietta in Georgia who might benefit even more from being the lead contractors on COST.

Cecily agreed that this time-honored approach was worth trying. But she was concerned that, with the present low US unemployment rate, the power of job creation to influence Congress was diminished. She needed to look for a Plan B and she was curious about what her other advisors would come up with.

When she returned to her office Cecily was annoyed to hear from her administrative assistant that she was scheduled to meet in an hour with the newly arrived NASA Congressional Interns. Every year it was her job to start them off with an uplifting pep talk mentioning the achievements of those who had gone before them.

Then she was expected to chat with each one, inquiring about their interests and priorities. Some of the women of color wanted to be assigned to a Latina or Black Congresswoman for example. Another might be hoping to work with a Congressional office dedicated to environmental impacts of space research. She had to do her best to match them with their dream destinations.

Cecily didn't think that this was the best use of her time as the Director of NASA's Space Science Division but her boss believed that it was important to demonstrate NASA's dedication to education. As she went through the tedious interviews matching interns to a list of Congressmen and women she came to one that changed her mind. "Maybe this is time well spent after all," she thought to herself.

The next intern settling herself into one of the chairs in Cecily's office was a striking brunette named Anna Grad. She had written that she was interested in understanding fundamentalist Christian opposition to science and was hoping to intern with a Congressman known for such opposition.

Cecily looked over Anna's credentials and saw that Philip had been her advisor and recommended her without reservation. Her first impulse was to contact him to check whether the pairing she had in mind for

Anna would be optimal. But she caught herself before she asked Anna's permission to do that. The glimmer of an idea that was forming in her mind might best be kept to herself, she thought.

Instead, she questioned Anna about her interesting but unusual request to work with a Congressman opposing science. "Do you have anyone specific in mind?" she asked. Anna answered that she did not know the records of the Congressmen well enough to narrow her choice to any individual. "But I would like to work with a delegate whose constituency backs active opposition to the teaching of evolution and global warming and supports Creationism," she answered.

Cecily didn't want to sound too eager so she took her time examining the descriptions of the priorities of various Congressmen before turning the document around on her desk so Anna could read the page describing the activities of Republican Jock White of Tennessee. "Here's probably your best option," she told Anna, modulating her voice for assurance and emphasis.

The description sounded reasonable to Anna. Besides his firm opposition to modern science, Jock had taken some positions on other issues that Anna agreed with. He opposed immigration across the Mexican border and wanted America out of foreign wars. His biography gave him as forty-five years old, six foot two inches tall, divorced with three children, owner of extended properties, and with hunting and fishing as his pastimes.

To relax Anna, Cecily couldn't help joking that she wouldn't mind meeting Jock herself. Anna's easy and untroubled smile at this somewhat risky pleasantry convinced Cecily that Anna might very well be the right girl for the job she had in mind.

Cecily told Anna that, if she agreed that working in White's office was what she wanted to do, she'd make the calls and arrangements. "Give me a day to sort all this out and I'll be in touch to confirm that you can start on the Hill on Wednesday morning," she assured her. "Meanwhile," she added, "take the day off and look around Washington tomorrow. You'll be pretty busy for the next year!"

Anna had toured Washington before so she spent Tuesday at the Library of Congress looking up material on Creationism, born again

Christianity, and on the activities of Jock White including, but not limited to, his Congressional record. She wanted to be prepared. She also emailed Philip, updating him on her arrival, thanking him for his support of her application and telling him that she hoped to return the favor by helping in some small way to ensure the success of COST.

Finally, Anna called her grandparents in St. Petersburg, Russia, before they went to bed. They were pleased to hear from her but not surprised at her call. Anna had always been a devoted granddaughter.

Chapter Twenty-One

Fund raising isn't as important a function of the Harvard Observatory's Director as it would be of the University's President. Still, Nancy had reason to believe that bringing in a new major donor would be a feather in her cap, so she was glad that Amanda shared her enthusiasm for recruiting Ivan. She looked forward to helping the University Development Office persuade him to join the Board of Overseers although there were still some steps to take, including the background check.

Nancy was a respected and well-liked scientist and had been appointed Director to restore calm to the Observatory after a dispute had blown up over controversial research carried out by the previous holder of the office. Nancy felt she was well qualified for the job. By mid-career she had done solid work on so-called peculiar stars but had discovered over the years that she enjoyed management more than research. She was bright and energetic and got along well with most of her colleagues, many of whom had difficult personalities.

Most had big egos—that went with being at Harvard. But a few took their intensely competitive pursuit of results to extremes. They were egomaniacs who showed little interest in anything but themselves; one had recently suffered a nervous breakdown when his research project failed to win funding. Others led a life of almost constant travel, circling the globe attending an interminable round of workshops, review panels, or conferences.

A few were borderline autistic loners whose interaction with other scientists, inasmuch as it existed, could be amiable but tended to be childish. Nancy considered Philip to be pretty normal and she had always gotten along easily with him. Like her, he had a spouse, children, and a home in the suburbs.

Sometimes Nancy wondered whether she was *too* normal to be really good at this occupation. She had moments of doubt as to whether she was just an imposter, a syndrome that an increasing number of her female colleagues were admitting to experiencing. She wondered why it was that the really big ideas always seemed to come from her male colleagues, even though she formally had the same background, opportunities and resources at her disposal.

It cheered her that astronomy had changed in her favor, she thought, over the course of her career. The trend seemed to be moving away from quirky individualists to people more like her, who could navigate the politics of large projects. A senior figure had once described her as someone who "rarely made a mistake." She wasn't sure what this virtue boded for scientific progress but she found it to be a more comfortable environment.

She was pleased that most astronomical institutions these days were run by women, but she wondered whether the enthusiasm for hiring them wasn't going too far. Her son would be graduating with a Ph.D. in physics in a few years and if universities were serious about achieving equal representation of women on the physics faculty, the job prospects for his generation of male physicists looked grim.

The work she most enjoyed these days included increasing participation in the American Astronomical Society's efforts on behalf of women and scientists from disadvantaged backgrounds. Organizing workshops and meetings enabled her to socialize with colleagues worldwide and she liked to travel as long as she could bring her family on occasion; even her youngest had come to like chicken biryani.

So when Amanda called saying that the background check on Ivan had shown no evidence of misbehavior Nancy looked forward to the next step. This would normally be an informal approach to Ivan by a Board member who knew him personally. The Board member might disclose how worthwhile of his own giving he found the Observatory to be.

The Chairman of the Overseers called Amanda a few days later saying that one of the Board members, who knew Ivan through business connections, had approached him about joining the Observatory Board. He was sorry to tell her that Ivan had turned it down, saying that he was interested in research at the Observatory but preferred to keep abreast of developments by maintaining direct contact with Philip.

When Amanda relayed this to Nancy, she immediately asked Philip to come over to her office when he was free for a chat. He knocked on her door an hour later and settled into an easy chair in her sunny office to discuss the situation. "Ivan already told me that he was open to helping support a truly novel project at the Observatory," Philip told her. "I hadn't gotten around to filling you in, but he just doesn't believe that he is suited to Board membership and has avoided it when asked previously." Disappointed as she was, Nancy could see that not everyone flourished in the Board environment.

"There might be an alternative way of keeping him involved," Philip proposed. "Perhaps he'd enjoy accompanying me on an observing run I have a few weeks from now at the Magellan Telescope," he added. "A trip to Chile together would give us a chance to exchange ideas on what kind of novel project he might be willing to support and just generally cement his connection to the Observatory."

Nancy had to admit that this seemed worth a try although she had a sinking feeling that now credit for attracting Ivan, if he did become a major donor, would probably go to Philip rather than to her. Still, the process was looking more like a marathon than a sprint, and there would be many twists and turns ahead. She had enough confidence in her skills to look hopefully to the future.

Chapter Twenty-Two

Several urgent phone messages had awaited Cecily on her return to Washington. She knew that, in these days of email and texts, the phone was reserved for issues best discussed without leaving a recorded trail.

She was not surprised that one was from her boss, the NASA Director, but another, from the Director of Aerospace at the Lockheed Corporation, was unusual. Two others were from the chairman of the Space Science Board, who routinely kept in touch with her, and the most recent from a Congressional staffer.

Cecily guessed that all of these calls were about the status of COST. Philip may have been untroubled by his decision to leave the project's leadership hanging, but she knew that his relaxed attitude wouldn't be shared by other concerned parties. She'd need to pull her thoughts together before returning any of these calls.

A remaining call was from Cordelia Brown, one of the two scientists who were embroiled in a dispute over ethics. Cecily frankly resented that Cordelia was dragging her into an imbroglio in which she doubted that she had any jurisdiction.

She wasn't aware of any precedent for Cordelia's request that another senior scientist's NASA funding be re-assessed because she had failed to alert colleagues to flaws in her work while accepting professional honors based on her results. Cecily found the evidence that Cordelia presented convincing but she felt uneasy about imposing penalties. She preferred to leave that to the National Academy of Sciences.

Cecily knew that professional ethics in science was a quagmire these days and it was hard to know where to draw the line. Complaints of plagiarism were so common that few paid any attention to them anymore. The basic problem was that there were relatively few well-posed problems to work on compared to the rapidly increasing number of scientists. So young researchers felt it was justified to borrow ideas from the literature without giving credit. Re-treading problems to re-solve them using more elaborate techniques that employed larger numbers of people was becoming disturbingly common.

Plenty of researchers behaved perfectly well but she continued to hear of breathtaking departures from decency that made her wonder what some members of the community thought they could get away with. One of her NASA grantees had complained recently that a proposal that he'd been sent to referee laid out a plan to steal an instrument that the referee himself had designed, and to continue its future development at a competing university laboratory!

Young people seemed to be particularly amenable to dubious practices. They were learning early that it was not unusual to write your own letters of recommendation, for instance, and have your Ph.D. advisor sign them. The advisors claimed to be too busy to behave ethically. Such practices were the thin edge of the wedge that led to more serious lapses later.

Communication of results had also changed. The rate of appearance used to permit colleagues to read what was new in their field and communicate about it by telephone or mail or over coffee at meetings.

Now, papers appeared at such a rate that few pretended to keep up with the literature. Results were discussed in the confusion of social media and it was difficult to follow developments without a web of connections through organs such as Facebook or Reddit. The new system was open to demagoguery and marketing skills had become as important as the quality of evidence. Once a sensational claim was picked up by the mass media it was difficult to counter because authors of popularizations weren't interested in retractions.

This all made it easier to tell Cordelia that she needed to study the issue in more detail. She didn't, but delay was a valuable tool in avoiding decisions. Cordelia knew that she was putting Cecily in a tight spot so she didn't deserve better. It also prepared Cecily for confrontation with bigger challenges before her.

Cecily knew what lay ahead. She had heard at the meeting in France that the potential international collaborators on COST were not enthusiastic. Philip's defection was also not a good sign, yet she would be expected to put the project in a positive light, whatever her feelings about it might be.

Cecily had been raised in an English family of modest means in a blue-collar area of Manchester. She remembered her family returning home after closing time at the pub; holding her mother's and younger sisters' hands while they walked ahead of their drunken father, who was staggering along behind, feeling his way along a red brick wall.

She had vowed to escape that tawdry life and she had. A scholarship had put her through a selective boarding school and another enabled her to study physics and math at Imperial College, eventually graduating with a Ph.D. in space physics. Then on to a post-doctoral fellowship at the University of London.

There she had been taken in hand by her elderly and influential supervisor who had valued her forthrightness and ability at data analysis. He had also fallen in love with her. Nothing untoward happened, but it helped her career. He took her along to foreign meetings and introduced her to many scientists of note.

Before long, Cecily had been appointed to a couple of influential committees. Even she felt a bit guilty about having been given precedence over male colleagues who had really accomplished more scientifically but didn't have the friends in high places that she had cultivated.

She acquitted herself well on the committees. Her hard work and clear presentations caught the eye of her future boss at NASA, who offered her an attractive position and assistance in getting a green card to work in the USA. Her father had died recently and she accepted on the proviso that she could arrange a card for her widowed mother as well.

She soon settled herself and her mother into a comfortable apartment in Virginia and had remained there ever since. Cecily didn't want to do anything that might put all this hard work in jeopardy, so when she called her boss back, she told him that she was looking forward to finding a strong replacement for Philip to spearhead the project.

She was confident, she also assured him, that private donors would enthusiastically fill any gaps in the funding left by any lack of agency support abroad. He told her that he was proud of her spirit and that she should let him handle the calls from Lockheed, the Space Science Board and the Congressional staffer.

On her way home that evening, Cecily stopped at a neighborhood bakery and bought her mum her favorite rum cake. It was chilly and raining outside but she turned on their gas fireplace and they spent a cosy evening together reminiscing about the bad old days in grimy Victoria Park.

Chapter Twenty-Three

Over the next few months Anna became accepted as part of Congressman White's team. He was away in Tennessee for a few weeks when she first arrived and at the beginning his staff wasn't sure what to make of her. None of them had ever met anyone with a Harvard degree and it all seemed as though she had dropped out of the sky.

But Anna told them how she had been brought up in part by deeply religious grandparents in Russia who also took the Old Testament literally and were skeptical of modern cosmology, evolution and other pieces of science that contradicted sacred teachings.

They liked her story of how in her first homework essay in the cosmology course at Harvard, she wrote that she doubted that the human brain was even capable of understanding the universe. When she told that anecdote at a bag lunch soon after her arrival her audience stamped their feet and nodded vigorously in agreement.

She told them that, in her year as an intern, she hoped to gain a better appreciation of the fundamentalist Christian outlook on science. She also wanted to seek out how the two outlooks might be reconciled. That brought consternation and dry coughing. "There is little hope of reconciliation, Anna," one of the Congressman's senior aides assured her. "The academic folks look down on us and have no interest in listening to our point of view."

Anna wasn't so pessimistic. "There are, after all, significant numbers of reputable scientists who are also devout Christians," she maintained. "Perhaps my work here should focus first on meeting some of those people and trying to learn from their insights." There was no dissent on that point so she set herself that objective in her research.

When Jock returned from Tennessee he invited Anna into his office and thanked her for her interest in his work. They chatted for a while and she described the initiative that she had taken in trying to understand whether reconciliation was possible. He smiled at her youthful exuberance and was impressed by her quick repartee. He also noticed that she was a very pretty girl. Her strikingly green, slightly slanted eyes reminded him of the eyes of a cougar that he had seen once while hunting for elk. The hunter in Jock was intrigued.

Anna's work kept her busy for the next few months and she got along well with Jock's other staffers. Most of them had degrees from Vanderbilt, Rice, or the University of Virginia and were as well informed as any of her Harvard friends. She socialized with them regularly because each of the Congressional office staffs kept to themselves and otherwise she wasn't aware of much nightlife in DC.

So she was surprised, just after Thanksgiving, to get an elegant printed invitation to a fancier party, a fund-raiser for research on Alzheimer's disease. Since this was a bipartisan issue there would be plenty of representation from both sides of the aisle. Most surprising to her was the affixed handwritten note from Jock saying: "I look forward to seeing you there, Anna."

Anna was curious about what all this meant and she bought an expensive little black dress and matching shoes for the affair. She had heard that Jock was well known amongst the Washington ladies and his divorce was the result of a highly publicized affair with a business partner's wife in Tennessee. He was a good-looking man, Anna agreed, but too smart, it was said, to get involved with any of the girls on Capitol Hill.

The night of the event was cold and rainy and she was glad that she'd had the foresight to have some of her winter clothes sent down by her mother. The address was in Georgetown and as the cab pulled up to a rather grand white-washed brick mansion she saw a flock of elegantly dressed people hurrying to the cheerfully lit open front door to get out of the weather. For a moment she was seized by a feeling that she might be out of her league, but her Harvard years soon kicked in. If you can cut a fine figure at a Fly Club party, you can pull anything off, she said to herself.

Inside there was the usual cacophony of Washington party voices—the male New England nasal twangs and fine baritone drawls of the genteel South interspersed with the bell-like clarity of female voices honed in years of board meetings. Anna ooched forward to where coats were being deposited and then moved on towards where she spotted the bar. She was glad she'd worn her highest heels; there were a lot of tall people. She didn't see anyone that she recognized and was not having much luck getting the bartender's attention when suddenly Jock appeared by her side and asked what she'd like to drink.

She asked for a glass of white wine, thinking that a slow start would create the best impression. Jock got quick attention from the bar for both of them and taking her gently by the elbow asked her to come along, he'd like to introduce her to some friends. Walking across the crowded room at his side, Anna noticed that they were attracting a lot of looks—the men smiling at both of them, fashionably dressed women appraising her. Even a few cameras flashed.

Jock told his friends, a small group of young men and women, that Anna had just graduated from Harvard and had joined his staff as an intern for a year. They seemed to be a few years older but interested in her work and her social experiences in Washington. They all worked for other Republican Congressmen or lobbyists and came from similar university backgrounds. Soon, one of the men asked her to dance and they wandered off to where loud music was playing down the hall.

The evening slipped by in pleasant reverie, and Jock checked in on her occasionally to see that she wasn't stranded or lacking a drink. She had switched to the Cosmo's she saw that other women were drinking and felt pretty cheerful, dancing with greater abandon with a succession of good-looking young men. Then, well past midnight Jock came over and asked whether he could drive her home; he didn't want her trying to find a cab or Uber at that late hour.

Anna was sufficiently sober to wonder whether this was a good idea given Jock's reputation, but searching for a ride did seem a lot more tedious so she said yes. He told her that he'd get the valet to bring the car around and he'd come to get her when it arrived. Anna found her way to the vestibule and picked up her coat but knew she wasn't feeling well.

Jock found her leaning against the wall in the entrance hall and helped her down the stairs into his black Mercedes sedan. Anna hoped she wouldn't be sick over the soft leather seats. The apartment in Foggy Bottom where she was staying wasn't far from Georgetown but by the time they arrived, Anna could barely stand up and Jock had to haul her out of the car. She gave him her keys to get into the building and up the elevator to her apartment.

Anna remembered now that her roommate was away for the weekend and realized that she had gotten herself into a potentially dangerous situation. She was alone, weak, and sick in her apartment with a man she scarcely knew. With her last energy she tried to thank Jock for the lift and ask him to leave, but he insisted on putting her down on her bed and taking off her shoes. Then Anna blacked out.

Chapter Twenty-Four

Ivan thanked Philip for his invitation to accompany him to Chile for his next run on the giant Magellan Telescope but he wasn't able to take a week off for that trip. As it happened, though, they both had some business to attend to separately in Los Angeles and at Caltech in the near future. So they agreed to meet in Pasadena for a night of observing at the famous Mt. Wilson Observatory located in the San Gabriel mountains nearby.

Philip was staying at the Athenaeum, the Caltech faculty club built in the early 1930s in the gracious style of old California, and Ivan drove up from LA to meet him for lunch there before heading up to the Observatory. Philip showed Ivan around the dark paneled library and reception room with its venerable easy chairs and into the large dining room. It looked like a men's club although it was hosting a faculty wives' lunch that day.

After lunch, Philip showed Ivan around the Caltech campus. "It's smaller than I imagined for such a powerhouse of science," Ivan commented. "Yes, there were only about 800 undergraduates and about as many grad students when I was here," Philip told him. "That was part of its attraction. You could find yourself standing in the cafeteria lunch line next to Nobel prize winners like the physicists Richard Feynman and Murray Gell-Mann or the biologist Max Delbrück.

"Part of my time here I was living in a little pool house I rented from a young couple nearby in the fancy adjoining area called San Marino," Philip continued. "They were kind to me and I invited them once to attend a special Caltech event—a public lecture by Richard Feynman on his hobby of deciphering Mayan hieroglyphics.

"The event was sold out with people standing outside the Beckman Auditorium watching it on TV," Philip recounted. "I'll never forget how Feynman affected that young woman. From the moment he came out on stage in his grey flannel slacks and white shirt with sleeves rolled up, she hardly stirred. I realized how genius can galvanize people."

Ivan had rented a Porsche in LA and offered to drive them both up to the Observatory. Philip didn't enjoy being a passenger in a fast car because in each of the two car accidents he'd experienced someone had been driving too fast to impress him. But he kept his feelings to himself and they headed out of town and up past Caltech's adjunct institution, the Jet Propulsion Laboratory, into the San Gabriels.

The Porsche's exhaust sang as they swooped up the curves of the scenic winding highway. After a while Philip relaxed, realizing that Ivan knew how to drive and wasn't pushing the famously tail-heavy 911 too fast into the curves. He wasn't looking forward, however, to the upper stretch, where the private road to Mt. Wilson branched off from the highway. There, the road narrowed and was not so well maintained, with big drops into the canyon below.

Ivan took Philip's warnings about the road into account, and they reached the Observatory safely and parked at the Monastery, as the astronomers' living quarters were called. Ivan was assigned a room and told to meet Philip in the kitchen after he'd freshened up after the trip. He was also told to shake out his shoes before putting them on in the morning to check for scorpions.

Over a cup of tea in the Monastery lounge Philip told Ivan about the history of the Observatory. He explained how George Ellery Hale from Chicago had pioneered the move of major observatories out of the vicinity of cities around the end of the 19th century and had chosen Mt. Wilson to build first a 60-inch aperture reflecting telescope and then the 100-inch. "He knew how to get private funding," Philip commented, smiling at Ivan.

"He had previously persuaded the Chicago streetcar baron, Charles Yerkes, to build the world's biggest refractor telescope at a site near that city. The 100-inch was finished in 1917 and it was the biggest in the world until the 200-inch Palomar telescope was finished in 1949.

"It was used by Walter Baade, a German astronomer working in the US, to get the first spectra of individual stars in the Andromeda Nebula during WWII when Los Angeles was blacked out and long photographic exposures were possible. These spectra proved conclusively that this fuzzy patch barely visible with the unaided eye in the autumn sky was a galaxy like ours, not just a pinwheel of glowing gas within our Milky Way.

"The 100-inch was also used by his colleague Horace Babcock in the late 1940s to discover magnetic fields on other stars besides the Sun. Their Pasadena colleague Olin Wilson used it in the 1960s to obtain stellar spectra showing that many other stars exhibited magnetic activity cycles like our sun's eleven-year sunspot cycle."

Philip had arranged to use the 60-inch to show Ivan some of the prettier objects in the sky that night. They would be joined by a few of the Observatory technicians and other staff who were looking forward to the event. They seldom had a chance to actually look through any of the telescopes they operated and maintained.

"Big professional telescopes are used to gather light and feed it into cameras and more specialized instruments like spectrographs and photometers," Philip went on. "These analyze the light into its colors to investigate the physical and chemical state of the glowing material far out in the universe."

Philip took Ivan into the 100-inch telescope dome to show him the instrument. "This was one of the biggest machines built by man at the time it was finished in 1917," he said. The large castings for the mountings of these giant instruments were dragged up the mountain road on trucks assisted by mule trains.

"The main bearing for the 100-inch, which carried most of its enormous weight, was a pool of liquid mercury. This enabled the very smooth motion required to accurately follow stars as the Earth rotated. But eventually they recognized that the fumes were responsible for the Mad Hatter–like mental deterioration experienced by the night observing assistant who was stationed near the bearing, so greater care was taken to prevent their escape.

"The light is gathered by the big mirror located at the bottom of the steel cage and fed to where it can be studied," he pointed out. "If you look up 60 feet to the top of the steel truss tube you can see the so-called Newtonian focus. To get at it, the observer stands on a little steel platform that moves with the dome as the telescope rotates during the night. The steady rotation of the huge telescope cancels out the rotation of the Earth once around in twenty-four hours. Otherwise, objects observed with the telescope would disappear from the field of view in a minute or two.

"Once I was observing on that platform using an electronic image tube that operated at a thousand volts. I had just focused it by crouching over and placing my eye to an eyepiece in contact to the back of the tube when it shorted out. A few moments earlier and the shock would have knocked me off the platform to the concrete floor 60 feet below. I was lucky.

"Even climbing up and down the ladder to the platform to develop photographic plates at three in the morning when you are groggy is dangerous. I don't think grad students would be allowed to do that in these litigious days, but forty years ago no one thought twice about it."

"This is like a shrine," commented Ivan as they emerged from the telescope's sepulchral surroundings of dark mahogany and black steel into the bright California sunlight.

"Yes; it projects a different vibe than modern observatories where the décor is stark white and the telescopes are usually painted cheerful colors like light blue or tropical green," agreed Philip. "The 100-inch is something of a national monument so its original state has been preserved. You'll find that the 60-inch that we will use tonight has been jazzed up with cheerier paint although it's older."

"It looks like a good night tonight; few clouds, no wind and no haze from forest fires at this time of the year. Dinner's at six, so you may want to doze off for a while now if we stay up late enough to catch Saturn and Jupiter."

Chapter Twenty-Five

At dinner in the Monastery dining room, Philip told Ivan about the old custom that the astronomer observing on the 100-inch telescope sat at the head of the table and the 60-inch telescope observer sat at the side. Also, the 60-inch observer was expected to carry the picnic basket containing the midnight meal for both astronomers up to the little building where that meal was enjoyed.

As they walked over to the telescope after dinner, Ivan noticed the aroma of the big pines and the stillness of the evening. A few of the brighter stars were already shining through the branches. "That stillness bodes well for the steadiness of the images we'll see in the telescope," Philip told him.

"The transparency up here about a mile above sea level is reasonably good when there aren't any forest fires around," he continued, "but the sky is bright because we are close to LA. The steadiness of the air, however, is exceptional. In fact Mt. Wilson was chosen over more exotic sites for interferometric observations at the highest attainable image sharpness. This good seeing, as astronomers call it, is actually the most elusive of the criteria used to rate the quality of observing sites and much effort is spent searching for it at distant locations like the peaks of Hawaiian volcanoes and in the Atacama desert of Chile."

Several of the mountain staff had gathered in the dome when Philip and Ivan arrived. The night observer was just setting the telescope on the early crescent moon, the first object on their schedule. It was at its highest above the horizon just after sunset.

The telescope's electric motors at first slewed it rapidly in the direction of the moon, then they stopped and the higher pitch of the fine pointing motors took over, moving the telescope the last few arc minutes onto the object. When the electric motors stopped buzzing Ivan felt a stillness descend on their hushed group. All was dark in the dome except for the moonlight pouring through the slit. He experienced a connection to the heavens that had eluded him after years of exposure to dramatic media reports on cosmic discoveries.

The operator moved the rolling platform over so he could climb up to the eyepiece. He'd never really seen the moon through a telescope except for glimpses through one of the antique telescopes he collected that were mounted on rickety tripods, and he was excited. "This is amazing!" he exclaimed when his eye adjusted to the bright light. Everyone in the dome laughed.

He stared, transfixed, at the hundreds of sharply imaged craters of all sizes, and the lunar mountains and canyons. "The moon is roughly a third of the Earth's size so everything you see there is roughly to scale," Philip informed him. "The highest mountains are about a third of the height of the Himalaya, for instance.

"When I was a graduate student," he continued, "it was still controversial whether some of those craters you are looking at were of volcanic origin. But now we know that they were all caused by impact of rocks of various sizes on the moon. No new impacts have been discovered since detailed observations became possible about a century and a half ago, but we know that our space neighborhood had a lot more space junk zipping around back billions of years ago when the moon was formed.

"Most of that junk has long since collided with the Earth or other solar system bodies," he added, "but there is still enough out there that an object big enough to destroy a city passes within the moon's distance from the Earth on average once a year. There have been near misses in the recent past; the Tunguska meteorite that grazed the atmosphere in 1908 and generated enough heat to ignite forests in Siberia is one. The big meteorite that struck near Chelyabinsk, Russia, in 2013 created more damage than was expected from models of such impacts. That was a sobering wake-up call that sent disaster specialists back to the drawing board.

"So the cosmos looks serene and friendly as we view the Man in the Moon smiling down on us, but celestial objects have the potential to inflict fatal damage on us here on Earth."

After everyone in the group had looked at the moon and the sky had darkened, Philip asked the operator to enter the coordinates for their second object, the Great Globular Cluster in the constellation Hercules. "It sounds crazy," said one of the telescope mechanics to Philip and Ivan as the telescope moved across the sky, "I've worked here for four years and this is the first time I've actually looked through one of the telescopes. Thanks for making this possible!"

"Well, if it makes you feel any better," said Ivan, "an Italian friend who builds Ferraris in Maranello tells me that he's never even ridden in one of the cars!"

When the telescope found its new object, Philip told Ivan to first sight along the tube just with his unaided eye and look for a small hazy patch amongst the sharper images of the stars. It took him a while to spot the object since the crescent moon had not set yet and the sky brightness was still somewhat elevated.

"What you are looking at there," Philip offered, "is the brightest example in the Northern sky of a so-called globular cluster, an amazing assembly of about a million stars jammed together much more densely than the stars in the sun's neighborhood."

Ivan saw what he meant when he looked into the eyepiece. "The field of view is just covered with hundreds of stars," he exclaimed.

Philip explained, "You are only seeing a small part of it because even with this lowest power eyepiece the telescope magnifies so much that the field of view is less than the size of the cluster.

"Globulars like this one also illustrate what seems to me to be a curious aspect of science. People have been observing them, and have known what they consist of, for over a century. Yet when I started teaching my astrophysics course at Harvard, I discovered that none of my colleagues working on galactic astronomy had given much thought to *why* they existed. Strange objects, really, if you think about it. Such a weird, senseless clumping of a million or so stars.

"It was strange to me that here, in plain view, were scores of these puzzling objects in our galaxy alone, yet few if any astronomers had cared to explain their existence. Science is replete with gaps like this, lacunae you might say, in our understanding that lie unnoticed all around us. Multitudes rush to study one topic while another, often adjacent and perhaps more interesting, lies fallow.

"Eventually globulars did attract attention when mysterious bursts of gamma radiation were spotted emanating from them a few years ago. But for many preceding decades they would have provided a curious young scientist an opportunity to answer a major question. So I have to smile when I hear people saying you need huge telescopes to discover anything worthwhile. What we need more is scientists who haven't lost the ability to sniff out what has been staring everyone else in the face, but gone unnoticed."

After everyone had viewed the cluster and, afterwards, the remarkable quadruple star system in the nearby constellation Lyra, Philip suggested that they adjourn for a midnight break and snack. "It'll be awhile before Jupiter and Saturn rise high enough to give us a good view," he explained.

"This is an amazing experience for me," Ivan exclaimed when they settled down with a sandwich and a coffee in the one room building mid-way between the 60-inch and 100-nch telescope domes.

"We're lucky to have good conditions," Philip responded. "Back in the days when most astronomy was done from ground-based telescopes it was considered satisfactory when you could work half the nights of an observing run. It's an unfortunate irony that the mountaintops astronomers favor at desert sites have the worst weather.

"Generally, in the US southwestern sites like the National Observatory at Kitt Peak near Tucson the summer is rainy season and the observatory is pretty much closed," he continued. "There is also a certain number of nights throughout the year lost due to high winds, forest fire ash, and instrument malfunction. Statistics show that on annual average, surprisingly, only about one third of the nights provide good data.

"It may sound grim, but you can get a lot done on a good night working from dusk to dawn," Philip pointed out. "I got most of the data for my five-year Ph.D. thesis project from about three nights of observing.

"I work with data from space-borne telescopes a lot now," Philip told Ivan, "It's less stressful, no question about it. I can work bankers' hours, get home for dinner, and avoid all the hassle of airports and travel arrangements. But there is something lost, for sure. There was a mystique of working in a place like this; I'm sure you can feel it. I even feel sorry for the younger generation, many of whom were brought up working in a cubicle with space data alone."

Ivan was interested in talking with the observatory staff members sitting nearby about their backgrounds and how they had come to work in such an out of the way location. There was plenty of time since Jupiter and Saturn were positioned well to the east of the objects that they had observed so far.

Eventually the 60-inch telescope operator came in announcing that he had a good view of Jupiter that they should come to see. So they all trooped out, treading cautiously while their eyes became accustomed to the pitch darkness now that the moon had set.

The seeing was really excellent now, after midnight, and even Philip was amazed at the detail visible in the giant planet's equatorial belts. "Notice how even the disks of the four major satellites of Jupiter are clearly visible," he told Ivan. "Those disks are about the size of our moon viewed at a distance of roughly a billion miles. So we are observing with a resolution of a few tenths of an arc second. That's remarkably good.

"It is also interesting that Jupiter is ten times the size of the Earth and there is some evidence that its huge gaseous envelope radiates a bit more light than it's reflecting from the sun. That extra radiation makes it almost a star, although its source would be heat extracted from contraction of the planet's enormous atmosphere. Jupiter is not massive enough to have sufficient temperatures and densities in its deepest core to trigger the nuclear reactions that produce the luminosity of the sun and other real stars."

Finally, the telescope was turned to Saturn, the night's pièce de résistance, as Philip freely admitted to Ivan. The view was absolutely splendid.

Even at the very high magnification of the 60-inch instrument, the image was rock steady.

"It looks like a picture postcard!" exclaimed Ivan, asking about the nature of the rings and of the razor-sharp divisions visible within them.

Philip explained that the rings consist of rocks ranging in size from centimeters to meters in diameter orbiting around the planet and susceptible to the gravitational tugging of Saturn's many moons. "The divisions arise from resonances between the orbital periods of the ring particles and the moons. The details of how this occurs were only worked out relatively recently, in part by Peter Goldreich here at Caltech." After all in the dome had viewed Saturn, Philip and Ivan thanked the staff and called it a night.

The next day, Ivan dropped Philip off on campus before continuing on to LA and home. "What you said about lacunae in science intrigues me," he told Philip, thanking him profusely for the memorable experience. "Maybe unsuspected paths towards understanding the universe are still staring us in the face, as you put it. Let's talk back in Cambridge on how I might be able to help discover them."

Philip spent a restful night at the Athenaeum. Next morning in the shower, he noticed the delicate lemon scent of the faculty club's soap and made a note to order some from Gilchrist & Soames when he returned home. It would have pleasant memories.

Chapter Twenty-Six

On his flight home Ivan thought about his experience at Mt. Wilson. He found the conversation that he'd had over the midnight lunch with one of the staff particularly intriguing.

Abe Reeves, the young electronics technician he'd spoken with, had been on the staff for about six years. He and his wife had decided to move out of the LA suburbs and home school their two children because they disagreed with the curricula the youngsters were facing in the local school system. "My wife and I are pretty moderate in our views," Abe told Ivan, "but when we learned what they were going to be teaching our eight year old daughter about sex in health class we decided that we needed to take a stand.

"I majored in philosophy and political science at UCLA and my wife was a fine arts major, but I couldn't find a job, even teaching high school. So I decided to learn electronics at night and landed a position up here. It suits us fine—my wife mainly teaches the kids and we love living up here out of the smog."

Ivan asked Abe how he and other mountain staff got along with the scientists. "The old timers tell me," he answered, "that the relationships seem to have changed. The astronomers they knew back in the 1950s and '60s were major personalities in the field who were identified with important advances in knowledge.

"People like Horace Babcock, who invented the magnetograph, was the first to chart the sun's magnetic fields outside of sunspots and discovered such fields on other stars. Olin Wilson, with his signature walrus mustache, showed that other stars have magnetic activity cycles like the sun. Bob Leighton was another remarkable figure; he used Hale's old 60-ft solar tower to discover the sun's five-minute oscillation.

"In more recent years the people we work with are fine folks, I'm sure, but I find it hard to identify them with any comparably important advances," he continued. "Also, from what I've heard, the social relationship has changed. Some of the young scientists nowadays treat me like a red-neck because I carry a deer rifle in my truck. I've tried talking with them about philosophy or politics but I get the sense that, unless I agree with all their views, they just tune me out."

"The old timers say that it didn't used to be that way," he continued. "It could be partly because Mt. Wilson isn't at the cutting edge of research anymore so we get a different breed here now. But personally I think it's part of the division of the country into opposing camps that the media are talking up all the time. People like me are viewed as yokels and the Ph.D. scientists set themselves up as holders of the real truth."

Ivan felt it might be better to steer the conversation to less sensitive ground but other staff seated nearby seemed to be interested in the subject so he asked Abe whether he had any ideas on how that division might be bridged.

"Well, for starters it would help to acknowledge that both camps have some work to do," he suggested. "Take for example what happened to President Larry Summers of Harvard when he suggested that there might be gender differences in math and science aptitude." Philip was silent; he'd kept quiet during that affair with Summers at Harvard, as had all his colleagues. He wasn't proud of it, but it hadn't seemed like a good time to fight for academic freedom when his case for tenure was under consideration.

"Beyond that," Abe continued, "without being presumptuous, I'd suggest that some of the ideas I studied as a philosophy major might help to address the divide in this country. Humans have always constructed myths to live by. Religion is based on myths for which there is really no evidence—only faith. So also, the liberals' belief that humans are perfectible is not based on any evidence, it's just another myth. The common thread in mankind's belief in myths is our need to believe."

There was silence when Abe finished. Everyone understood the import of what he was saying. Philip recalled his conversation in Maine where he had suggested to Reverend Larry that, in cosmology, it was presently pretty much a matter of faith whether you believed in 10^{37} or more parallel Universes or in a Divine Creator.

Ivan found that intriguing and thought about it a lot after they retired that night. He wanted to understand the concept of faith. People used the word all the time but if you thought about it, it seemed to go against human nature to adhere to beliefs which had no basis in fact or even contradicted well established knowledge. So what, really, was faith?

Understanding our puzzling need for faith, he felt, might help to address the increased skepticism directed towards science these days. More generally, it might even help to bridge the widening chasm that divided the political left and right in the Western world.

Chapter Twenty-Seven

Anna woke up next morning naked and sore, and she quickly realized that Jock had not left after putting her to bed and removing her shoes. But she needed some time to admit to herself that the eminent Congressman and friendly mentor Jock White had raped her while she was blacked out. A flood of emotions gripped her. Anger at him for betraying her trust and at herself for being so gullible.

At first she just lay in bed letting all these feelings roll over her, but she was too strong and angry to let this go on for very long. She got up, had a shower, dressed, and made herself some coffee. She didn't know anyone in DC well enough to confide in and ask for advice, and calling her mother was out of the question. Certainly not anyone at the office. Then she thought of Cecily who was after all the person who had most experience with situations that might confront her interns.

She hadn't kept in touch with Cecily much except for a couple of brief updates on her progress. When she placed the call she was told that Cecily was in meetings all day and would try to get back to her later in the afternoon. Anna's emotions got the better of her and she blurted out: "Tell her it's really urgent—I've been raped!" On hearing this, the receptionist came to life and assured her that Cecily would call back within five minutes.

She was true to her word, and Anna told her the story of how Jock had invited her to the fund raiser and what had happened when he took her home. Cecily got right to the point and told her to stay put and absolutely not communicate with anyone until she got back to her.

"I'm sending my assistant down immediately to pick you up and take you to a clinic for examination," she told Anna forcefully. Anna protested that this wasn't necessary but Cecily insisted. "Just stay in your apartment, and she'll call you from the lobby when she arrives. I repeat—don't talk to anyone before I get back to you. Not even to your parents."

After she got off the phone and sent her assistant on this errand, Cecily sat and stared out of her office window for a few minutes thinking about the implications of this development. She called down to the meeting she'd been in and excused herself for an hour, citing an emergency.

Cecily was a good and compassionate woman and she felt badly for Anna, but she couldn't help but let her mind roam further afield to the wider implications of this event. Handled skillfully this could, if not destroy Congressman White's career, at least cripple his credibility in opposing the COST project.

Cecily was not proud of thinking politically instead of focusing on poor Anna's situation, but she assuaged her guilt by wondering why girls as bright as Anna seem unable to avoid passing out drunk after parties. She thought to herself that, if she had a daughter, she'd make damned sure that she took responsibility for her own well-being. Getting drunk regularly and putting all the onus on the men was just crazy.

Cecily knew that she'd need to bring her boss in on this if the matter was to be brought to the attention of Capitol Hill and the media. He'd want to bring in NASA counsel to examine the legal ramifications so she decided to wait until her assistant returned with Anna from the clinic and they'd had time to discuss this a bit more calmly. She also decided to postpone telling Philip. She knew that he felt protective of Anna and would be very upset if he found out through the media, but she'd call him later.

The clinical exam confirmed the rape but assured Anna that physically she was not harmed. Cecily sat with Anna in her office for a while going over the facts and then invited her to lunch in the NASA cafeteria. As they were passing through the line to pay, the jovial cashier lady gave Anna a big smile and exclaimed loudly for all in line to hear: "Congratulations, hon, I seen yo' picture in the paper today with that han'sum Congressman!"

Anna was taken aback and tried to deny it saying it must have been someone else, but the cashier insisted: "Go buy yourself that newspaper for sale down there, girl, an' you'll see fo' yo'self!"

Cecily paid for Anna and bought a paper from the stand in the cafeteria and sure enough, on the social page Anna saw a half-page picture of herself smiling on Jock's arm at last night's party. The caption read: "Who's Jock's new friend?"

"You are famous," said Cecily quietly. But she knew that this moved the schedule up considerably. If the media were onto Anna and Jock, NASA needed to get this to the appropriate channels on the Hill before some enterprising reporter dug up the facts and splashed them all over the national news.

She called up to her boss's office and requested five minutes of his time right away. His secretary told her to come right up, and Cecily decided to bring Anna along for emphasis. After a brief introduction Cecily showed her boss the picture of Anna and Jock and filled in the details. She asked for permission to put in a formal complaint against Jock White to the Congressional office charged with disciplinary matters. She also suggested that the NASA press office be brought in to handle any contacts with the media.

Her boss agreed that the rape required action by federal law and that the Hill must be alerted. He called down to the NASA legal office and asked his Chief Counsel to join them immediately. He confirmed that they had no choice but to report it and suggested that they let the Congressional press office handle the interface with the media. He asked Anna to write up the facts in Cecily's office so the memo could be sent to them as soon as they were requested.

He also stressed to Anna the need to lie low and not communicate with anyone. Cecily asked her assistant to drive Anna back to her apartment. If even the cashier in the cafeteria recognized her, she couldn't be too careful.

When Anna returned to her apartment, she was shocked that even that precaution had not been sufficient. There were several phone messages waiting for her from various media contacts. She ignored them and did not pick up the phone all afternoon. She took another shower and went to bed for a badly needed nap. When she awoke, she ordered Chinese take-out and watched some movies well into the night.

Next morning Anna checked her email and saw a message from the office. It was a formal document written on the Congressman's letterhead. It said basically that documents had been found on her computer which showed that she had obtained the internship in Congressman White's office under false pretences with the intention of entrapping him.

It went on to warn her that any attempt to make allegations against the Congressman would be met with the most serious legal consequences. Finally, it informed her that she was no longer allowed on the premises of the Congressman's office and she should consider her work there terminated.

Anna was shocked. The event last night had been traumatic, but she could handle it. Nothing had happened that wasn't within the realm of experience for her or at least for other girls she knew in college. Friends had gotten drunk at parties and found themselves in bed with a stranger next morning. It wasn't pretty but they got over it. To be caught in an intrigue with national consequences was on a different level.

Anna called Cecily's personal phone as soon as she had absorbed the contents of the letter. She told her that she had no idea what documents Jock's office was referring to and didn't have any idea how they could have construed entrapment. Cecily told her that she didn't know what they were talking about either, but that she'd need to consult her boss and counsel before proceeding.

She supposed that the Congressman's office had been alerted as soon as the complaint was lodged yesterday and that they had grabbed her office laptop off her desk and spent the night poring over its contents. She repeated her instruction to stay out of sight even if it was tedious. She offered to come by to have a take-out dinner that evening. Anna thanked her but said she was fine reading a favorite Russian novel, Mikhail Lermontov's *A Hero of Our Time*, on her Kindle.

Anna also searched through all the documents on her personal laptop for anything that might possibly be used to construe a plot to entrap Jock. She came up with nothing on her first pass. Then, while she was thinking of people whom this affair might touch, she remembered the email she'd sent to Philip on the day before she started work.

She had been issued a laptop by NASA for her office use that day and had used it to send the note to him. She couldn't recall the exact words but it seemed to her that she had, in fact, said that she hoped to help COST in some small way during her internship. That email might not look good in court, she imagined, given that Philip, a key Co-Investigator on COST, had recommended her for the internship. Also that Cecily, whose office was to fly the telescope, was in close contact with Philip and had specifically sent Anna to intern with Jock, a sworn opponent of COST.

Chapter Twenty-Eight

Cecily had stayed in touch with Philip after they returned from France because she was hoping for some advice on who might lead the new telescope project. He wasn't able to suggest any likely candidates at Harvard or MIT. Astronomers who had experience with instrumentation, clout in the community, and the people skills to pull off a project of this scale were hard to find. She realized that she'd need to look farther afield.

Members of her NASA Advisory Committee suggested she look outside of academia, to scientists at government laboratories like the Applied Physics Laboratory associated with Johns Hopkins University, the Smithsonian Astrophysical Observatory that shared space with the Harvard College Observatory, or the Jet Propulsion Laboratory that belonged to Caltech.

The personnel at these labs really had more experience in handling large projects than did academic staff. The downside was that it was harder to find the star quality that an internationally acclaimed professor from a top university would confer on the project. Cecily needed a star who could sell a project of this scale at the Congressional level. The star system is as important in Big Science as it is in sports. Everyone wanted to see Wayne Gretzky play hockey. The next-best guys played to a lot of empty seats.

As often happens, challenging problems can yield to new insights from unexpected quarters. A few days before Cecily was scheduled to meet with her highest-level advisory group, the Space Science Board, she was notified that one of the key members would not be present because he had scheduled a scuba diving trip to Belize on that date.

This irresponsibility angered her; he knew that he was the Board member with the most experience and clout in astronomy and that she needed him at this meeting. The Board otherwise consisted mainly of upper atmospheric physicists and chemists, and experts in various aspects of solar and terrestrial plasma physics. She privately cursed his irreverence and sheer cussedness in prancing off like this.

She was in no mood to accept advice from this renegade but time was short so she agreed to the replacement he suggested, a relatively young fellow from Ball Aerospace named Brian Oglethorpe. It was a departure to have a representative from private industry on the SSB, but he came with strong recommendations, so she gave this substitution her nod.

Even after years of experience, Cecily dreaded these advisory committee meetings. The agenda might look routine and harmless but only seldom did the two- or three-day events pass without some unexpected flap that left egos bruised and turned friends into life-long enemies.

Of course it wasn't all bad. Several of the members, when she had arrived on the job, had been kind to her and given her good advice. Also, some of the members seemed to enjoy the stress and the sparring and once in a while fresh faces appeared and injected some welcome positive energy into the deliberations. She was looking forward to meeting Dr. Oglethorpe.

The traffic was bad coming in on the morning of the first session, and she was running late. She swept into the meeting room and unpacked her laptop noting with some satisfaction out of the corner of her eye that these fusty academics still appreciated the sight of an attractive woman. Even the female board members seemed to be cheered by her well-cut grey suit and hint of French perfume. They understood.

As she expected, the members had already chosen their seats, although some were still conferring by the coffee and pastries. A seemingly innocuous formality, this choosing may seem, but she remembered reading in Henry Kissinger's memoirs how the opposing delegations at the Vietnam peace talks had spent the first several days wrangling over the shape of the table to be used. Here also, she had noted that certain personalities preferred to occupy seats near her head of the table whereas others made a point of sitting nearer the opposite end, facing her.

Cecily noticed the new face, an African-American scientist, who she assumed was Dr. Oglethorpe, already working on his laptop. She walked over to greet him. He was a handsome man of somewhat more than average height with a ready smile. She welcomed him to the Board saying he'd be an important asset in their deliberations over the next three days.

She then took her place at the head of the table and called the meeting to order, welcoming everyone to Washington and hoping they'd had pleasant travels. The main item on the agenda was a review of the projects called for in the ongoing Ten-Year Plan for Astronomy. Astronomers had held to such a decadal planning exercise since it was introduced in the 1960s. The first few iterations had proven very successful in providing the agencies and Congress with an orderly set of priorities to be carried out in the next ten years. It had been so successful, in fact, that the model was adopted by other areas of science besides astronomy.

Unfortunately the projects became too ambitious to carry out in a decade and unfinished ones began to clog the pipeline. The latest was the James Webb Space Telescope whose delays and massive over-runs used up resources originally earmarked for several smaller, high priority projects. The overruns also threatened funds for analysis of data from previous missions as well as support for theoretical studies. The Board was now tasked with deciding whether the next Ten-Year planning exercise should be delayed until the backlog had cleared, or even whether the JWST should be cancelled to enable other projects to proceed.

The meeting began with a series of thirty-minute presentations on the projects in the pipeline, each given by a Board member who had been assigned that project a month ago. The presenters had no personal stake in the projects they described so the reports were dispassionate and there were few questions. They were able to keep to the schedule and with a short break for coffee got through the lot by lunchtime.

Cecily asked Brian whether he'd like to join her for lunch in the cafeteria. A few others soon rounded out the group. She inquired about his background and learned that he was from Alabama and had earned a Ph.D. in astronomy at the University of Texas in Austin. He had then gone on to get a Master's degree in aeronautical engineering from Rice University in Houston. He had found, during his Ph.D. studies in astronomy, that he was more interested in space hardware and advanced instrumentation. Ball Aerospace in Boulder, Colorado, hired him straight out of Rice. They were in charge of the detectors used on the JWST and he had risen rapidly through the ranks in the easy-going Rocky Mountain atmosphere they offered.

Brian had given a solid presentation that morning without causing any controversy. But at lunch he surprised Cecily by launching into a forceful exposition of his views on the shortcomings of the Ten-Year planning exercise. "It seems to me that, if we are focused on realizing COST, we need to adopt a totally different approach to making our case to Congress," he pronounced. "What we are doing here is just re-arranging the deck chairs on the Titanic," he continued. "We're never going to raise the $20–30 billion we need by running out the paradigm that worked for getting a few hundred million twenty years ago," he assured them.

"One of our Program Managers at Ball, Dave Powers, showed me his notes from the visioning meeting in Normandy," Brian told the group. "It's pretty clear that no compelling scientific reason has been identified for building COST," he continued. "I doubt that Congress will get excited by an open-ended fishing expedition when they could buy a fleet of nuclear subs for the same money. We need to move away from the science lobby and over to the much more powerful aerospace lobby if we want to raise those kinds of funds.

"Essentially," he continued, "let's stop pretending we are selling science and admit that we are in the business of developing new aerospace and military technology, providing employment for tens of thousands with a possible upside of discovering some exciting new stuff about the universe. There's no limit to the funding we can access that way."

Cecily listened to him carefully and by the time they had finished dessert and coffee she had a warm feeling that Brian, with his fresh outlook, aerospace connections, and minority status might be her man to fill the leadership gap left by Philip. She also looked forward to having dinner with him. She found his sonorous voice to be almost hypnotic.

Chapter Twenty-Nine

Brian and Cecily got along well over dinner in a quiet Indian restaurant that she had discovered. So well, in fact, that Cecily accepted Brian's invitation for a drink afterwards at the tony little inn in Foggy Bottom where the National Academy generally put up committee members. Cecily agreed that it would be wiser to have their drinks brought up to Brian's room to avoid encountering other SSB members in the hotel bar.

One drink led to another and Cecily never made it home to Virginia and her mother that night. She couldn't suppress some guilt at such fraternizing with a government contractor, but she slept well nevertheless, assuring herself that this sort of thing happened all the time and there was no way it would sway her judgement in contracting out COST if it ever came to that.

Any remaining feelings of guilt were banished when Cecily noticed in the mirror next morning that her short blonde hair had a new bounce, her eyes seemed bluer and her English peaches and cream complexion seemed even peachier. She wondered whether this glow would last the three months to the next SSB meeting or whether she might need to arrange a site visit to Ball Brothers before then, to see how they were making out on the JWST detectors.

Cecily had been married in graduate school to a young colleague, a dapper Scotsman who was congenial and easy going. But it became apparent after graduation that they were on different career paths and when she was offered the NASA job in Washington they decided that he wouldn't be following along. They were divorced soon after; he found someone in London who was less career driven, and she settled into her NASA life that required a lot of travel and long office hours.

The job was stimulating and it was fulfilling her career goals but she gradually realized that competition for eligible men in Washington was fierce. There were many pretty, younger women hunting for the few unmarried men with career prospects that impressed her. She had tried the alternative of marrying someone on a slower track, but that had not been a success. So here she was in her early forties, still living with her mother.

Brian was a few years older, also divorced, but with two daughters. Still, they had much in common. Additionally, she reminded herself driving back to the meeting from Foggy Bottom next morning, he might be the man she needed to replace Philip in leading the COST project. He wasn't the star academic that she had been shooting for, but she thought that, by backing him up with some wizard advisors, she could make it work.

The second day of the meeting began with presentations by chairs of other NASA and NSF advisory committees, including the co-chairs of the most recent Decadal Review. Cecily was impressed by the thoughtfulness and diligence that had been invested in these reports. There were differences in detail but the general gist was of support for keeping the Decadal Review process on schedule, but scaling back new projects until the JWST was operational.

The COST project was seen, for the moment, as a challenge on the horizon not yet ready for more than visioning such as provided by the meeting in France. In the opinion of some, astrophysics was headed for the same dilemma encountered by particle physics—there was no clear path ahead and costs were reaching prohibitive levels. Both fields seemed to be facing a wall.

In the afternoon the SSB met in executive session to discuss the material and opinions that had been presented. Brian repeated the case that he had made at lunch the day before. He outlined how missions of the size and price of COST could piggyback on requirements to provide employment in aerospace and new technology for commercial and military applications.

Any new insights that they might provide into astrophysical questions like the nature of the universe or the existence of extra-terrestrial life would be welcomed, of course. But the science would no longer be required to bear the weight of justifying the funding required to build the colossal next-generation space borne instruments.

As Brian spoke, Cecily tried to visualize how he would do in confrontations with pugnacious Congressmen who did not automatically sympathize with anything that scientists dreamed up. Some, she knew, were openly hostile towards science and would challenge COST at every opportunity. But Brian's arguments had a convincing ring to them that the usual pleas for more money from the science community lacked. They were arguments that any Congressman could take back to constituents and please both conservatives and liberals.

She was also impressed by his delivery. He was confident and forceful without getting strident and antagonistic. He was easy to like. She remembered something she'd read in a novel by the Czech author Milan Kundera that a man's voice determines, more than any other trait, the impression that he creates. Listening to Brian, she couldn't agree more.

There was still the impact of Brian's minority status to consider. These days an initiative projecting the image of diversity would only, it seemed to Cecily, be strengthened. Some might argue that minority leadership could be problematical because any criticism of the project immediately dragged in racial or gender overtones. But such considerations were not carrying the day in government circles these days. On balance, she believed that COST would benefit from a champion like Brian.

After the meeting closed on Friday afternoon, Cecily and Brian agreed to meet at Reagan Airport where he would return his rental car and they would continue to her place in Virginia so she could pick up some outdoor clothes. He had planned a weekend visit to Williamsburg and the Chesapeake Bay East Shore before returning to Colorado, and Cecily was happy to come along when he invited her on the spur of the moment.

As they left the airport, Cecily cheerfully let Brian take over the wheel of her new Buick. She introduced him to her mother who was very accepting and happy that Cecily had a man in her life—it didn't seem natural for a girl as attractive as her daughter to just be working all the time.

Chapter Thirty

Anna called Cecily and told her about the email she remembered sending to Philip on her office laptop. They agreed that this was likely to be the evidence that Jock's office was citing. When Cecily discussed this new development with her boss and NASA counsel she was told that it would be best for Anna to retract her accusation before this matter went any farther.

Whatever damage this accusation might cause to Jock, it would be minor compared to the fallout for NASA and COST if even a hint of entrapment of a popular Congressman by a Federal agency surfaced in court and in the national media.

Cecily asked Anna for permission to come by right away to discuss developments. She brought some documents with her to make the withdrawal of her complaint official. Essentially she had her attest that whatever occurred was consensual and her previous allegations were the result of not properly understanding the law on such matters. She scanned them immediately on equipment that she had brought to Anna's apartment and sent them to the NASA counsel who conveyed them to the Hill.

Cecily suggested that now that the matter had been defused they could have dinner together in a little Moroccan restaurant located at walking distance over the bridge in Georgetown. Anna consented gladly; it was a fine evening and she happily discovered that she really liked lamb tagine.

Over dinner Cecily suggested that Anna might benefit from a change of scenery and that, since the internship description was pretty broad, she might want to get out of the country for a while to avoid press attention and clear her head. She asked whether there was any place that Anna wanted to visit in particular, say for a few months. Cecily offered to contribute some NASA discretionary funds to help with travel expenses. Anna thanked her and told her she'd give it serious consideration.

That night, Anna thought about how Lermontov's hero, a dashing Russian cavalry officer, had become besotted with a beautiful Circassian chieftain's daughter, and had carried her off. But he misused her and in the end neither had met a happy end. Anna wondered what fate had in store for Jock.

On her drive home that night Cecily mulled over the drama of the past few days. Information from Brian was that Plan A to fund COST over fundamentalist opposition was making slower progress than he had hoped. The problem, as she had predicted, was continuing strong employment figures.

Now, her Plan B had also misfired. But here she was conflicted. She admitted to herself that she was actually relieved that the scheme had not gone any further. Like Anna, she was starting to feel that she had put in motion something that was spiraling out of her control and could have nasty consequences for her and for NASA. For certain, her mother would have been aghast if she had ever discovered what her Cecily had been up to.

Next morning Cecily got a call from Anna saying that she wanted to accept her offer and that she'd like most to spend the remaining time on her internship in St. Petersburg with her grandparents. She knew some of the astronomers at the famous Pulkovo Observatory in that city and could spend some of her time there, if Cecily thought that would be helpful in fulfilling her obligations to NASA. She also told Cecily that, after her experience in Washington, she had decided that she would not pursue astronomy in graduate school. She planned to take the LSAT exams and apply to law schools.

Cecily smiled at this ending to a stressful episode. She was happy for Anna that she was looking forward to a bright future. But unfortunately her own problem of fending off Jock White's opposition to COST had not gone away. She needed to find a robust backup plan.

Chapter Thirty-One

A few months after Philip had hosted Ivan at Mt. Wilson he ran into Nancy Stein at one of the teas held before colloquia at the Center for Astrophysics. She asked whether he'd heard from Ivan since that trip. "You might want to reach out to him casually so we don't lose touch," she suggested. "We are able to contribute to your entertainment costs; I have some discretionary funds for development activities," she added with a conspiratorial smile.

Philip cringed whenever the expression "reaching out" cropped up, but he admitted that he'd been preoccupied with other matters and agreed that he should inquire whether Ivan might come back to the CFA for a second visit, perhaps to meet some of the other faculty.

But driving home that evening Philip thought that he had a better idea. He remembered an invitation to give a review paper at a cosmology meeting to be held in Prague in two months. It occurred to him that this would be a good excuse to contact Ivan. If Ivan spent as much time in the Czech Republic as he had said maybe he'd be there around the dates of the conference and might be willing to show Emily and him around. Emily had been kidding him for years that they were the only Cantabrigians who hadn't been to Prague yet.

Philip called Ivan that evening and Maika told him that he was abroad on business but would get back to him when he returned in a couple of days. She was enthusiastic about the opportunity to meet with him and Emily in her hometown in the spring when the city was full of music and at its best.

When Ivan returned to Cambridge he contacted Philip and enthusiastically endorsed meeting in Prague. He had been planning to go over about that time to transact some business and also to look over an antique car that he was being offered. He explained to Philip that, although he wasn't really a collector, this particular car had special significance for him.

"My father had always been fascinated with the racing cars built in the 1920s and '30s by the Frenchman of Italian descent, Ettore Bugatti," he said. "This example of a model that my father particularly admired has recently been found stored in the outbuildings of a renaissance palace near Prague. A business associate alerted me to the opportunity and urged me to come soon," he added.

Philip accepted the invitation to give the talk and they arranged their schedules to enable him to spend a few days at the conference while Ivan attended to his business affairs. Maika would spend that time visiting relatives and Emily would fly in a few days later to join them.

Ivan wanted them to come along on the viewing of the Bugatti. "I need your opinion before I decide," he told them with mock seriousness. "Furthermore, the palace where it is located is worth a trip in its own right. It has some connections that might interest you," he added. Emily told Philip that he should read up on Bugattis so they didn't come off as completely ignorant. She planned to learn something about Czech history for her part.

Philip thought that besides the opportunity this trip would afford for courting a potential donor, it would also be good for him and Emily to get a change of scenery. After a year of frenetic applications and interviewing, their daughter Becky was safely placed at Wellesley College near Boston. Her first term had gone well and she seemed to be prospering in its steadfastly but not aggressively female environment.

William, their son, had managed to get into a competitive prep school in New Hampshire for next fall, but they continued to worry about him. He seemed morose and still had trouble making friends. Philip had difficulty understanding how a boy as gifted and fortunate in his upbringing as William could seem so uninterested in life.

His diagnosis was that William was a victim, like many of his peers from well-to-do families, of having been given too much. So many options frustrated and ultimately angered him. Emily's parents cautioned them against holding up Becky to William as a paragon of virtue and success—they saw that as part of William's frustration. Whatever the explanation, William had graduated from tantrums when he was little to general despair as he grew older.

Philip and Emily made a point of visiting him often and worked hard at putting a positive light on any successes that he achieved academically or in extracurricular activities. But they saw little long-term progress and their differences of opinion on how to address William's situation were putting increasing stress on their marriage. Philip urged Emily to let go and give William space to make it on his own. She found that too difficult. A trip to Prague seemed to both of them like it might provide a refreshing break.

Chapter Thirty-Two

Philip's flight to Prague was through Munich and he just made the connection. He wondered whether the airport had been enlarged since he'd been there a few years ago or whether he just tired more easily now. The distances between gates seemed longer.

After they took off for Prague the passenger in the aisle seat next to him, a tweedy, ruddy faced gentleman, asked him cheerfully whether he knew that air plane hijacking was a Czech specialty. "In 1950 three DC 3 airliners were hijacked simultaneously by their Czech airlines crews and flown to Germany," he informed him triumphantly. Philip smiled weakly but thought that this kind of humor would put you into detention on an airline flight back home.

"You know," his new friend added with a smile, "we Czechs have an unfortunate habit of inventing bad things. It is true that our talented chemists came up with soft contact lenses which are beneficial but they also invented Semtex, the convenient stick-on explosive that was widely used by terrorists to blow up airliners. Our President at that time, Vaclav Havel, wisely stopped its production soon after he came to office. He wanted our country to be known for its happier contributions, like world-class beer, hockey, and super-models!"

Philip's attention gradually turned to the view out the window as the plane descended toward Prague. The flight became bumpier as they passed through some dark cumulus clouds. Shafts of sunlight here and there illuminated orderly fields below. The raindrops scurrying along the plexiglass augured a rainy introduction to Prague.

His aisle mate noticed that Philip gripped the seat handle when the plane dropped suddenly here and there between updrafts. "Don't worry," he assured Philip, "Czech pilots are very competent. I like to remind my English friends that the ace who downed the most German planes at the Battle of Britain wasn't some Old Etonian. It was Captain Josef František, a Czech pilot who flew a Hawker Hurricane with the RAF. He was credited with seventeen kills before running out of gas one fine day and ditching in the Channel to end his brilliant career.

"He had a novel approach. He didn't get along with other pilots so he flew alone and waited for the Germans to be returning home when they had little gas left for dog fights. This may have helped him compile these impressive statistics," he added with a wink.

Philip was relieved when the plane finally bounced onto the tarmac and he was able to gather his belongings and bid adieu to his garrulous fellow traveller. But as they walked towards the terminal through the drizzle his informant still took the opportunity to add: "As you are admiring Prague architecture during your visit keep in mind our country's most questionable contribution to humanity: Protestantism." Then he waved goodbye and disappeared into the throng. Philip made a note to ask Ivan what this puzzling comment meant, but for the moment he became fully engrossed in arrival formalities and collecting his luggage.

Ivan had arranged for Philip to stay at the Villa Lanna, a neo-classical palazzo that had once belonged to his family. It now was the property of the Czech National Academy and visiting scholars were often housed there. It wasn't in the center of town, but its location in a leafy suburb amongst embassies and other stately homes would help get him some peaceful sleep. It was a short walk to a direct subway ride straight into the heart of town.

Philip hailed a taxi and was soon on his way through the fitful rain into the city. The outskirts looked much like most European cities these days; uninspiring boulevards fronted by office towers of various descriptions and apartment blocks that he was grateful he didn't need to live in. Modernity intruded here and there but it didn't take much imagination to still see Harry Lime, from Orson Welles' famous film *The Third Man* set in post–WWII Vienna, pop out of a side alley. The grey skies didn't help.

But eventually the cab turned into a tree-lined street with the beginnings of spring greenery softening the surroundings of cobblestone pavement and grey masonry houses. Red geraniums brightened window boxes here and there. A few blocks later the apartment buildings were replaced by large villas with extensive walled gardens, some of 1930s Bauhaus style, others dating back to the 19th century.

The cab stopped in front of an imposing villa larger than Philip had expected. The driver helped Philip carry his bags through a formal, varnished entrance door into the reception area, a large marble-floored hall with a curving grand staircase winding up along the wall towards the floor above. After checking in Philip was ushered up this staircase down an expansive hallway and through a formal sitting room to a lavishly proportioned bedroom with ceilings that must have been fifteen feet high. Ivan's ancestors certainly lived well, thought Philip to himself.

The valet drew open the heavy drapes letting some welcome natural light into the room and saw to some other final arrangements. He also gave Philip an envelope that had been left for him at the reception desk. Opening it, he read a note from Ivan and Maika welcoming him to Prague and to the Villa Lanna and saying they were out of town for a few days but would return the day after tomorrow after his conference wound down. They were looking forward to meeting up when Emily flew in that day.

He thanked the valet and asked to not be disturbed. The fluffy eiderdown often seen in Central European bedrooms was so comfortable and smelled so sweet that he fell asleep almost immediately.

Chapter Thirty-Three

When Philip was awakened by his alarm it took him a while to recall where he was. Daylight filtering through the wispy curtains on the tall windows illuminated a few colorful expressionist paintings on the room's high walls. The light-colored natural wood furniture was of an appealing pre-war art deco design. He washed in the adjacent bathroom attractively tiled in black and white mosaic with chrome plumbing fixtures again in well-polished 1930s designs.

He was glad that he had brought a pair of brown cordovan leather brogues in case it was too chilly to wear his comfortable boat shoes. Actually, come to think of it, he'd not feel right walking around this house in boat shoes anyway. He checked for effect in the bedroom's full-length mirror before locking the door behind him and descending along the winding staircase to ask the receptionist for a cab.

The driver recognized the destination instantly and with a deferential glance at his passenger started the meter as they pulled off into the spring evening. The cab rattled over cobblestoned streets, climbing upward through many twists and turns. They passed out of the section of villas past what looked like a walled military compound and eventually halted at a sentry post where a soldier peered into the car and allowed them through. Then they passed along an alley between high baroque yellow stucco walls and crossed a deep moat into a large courtyard ringed with amazing renaissance and baroque palaces.

Philip leaned forward to ask the driver where they were. "Hradčany Castle, sir," he was told. Philip wished that he'd read over his Fodor's guide instead of watching a movie on the flight over from Boston. After rattling across the cobblestone courtyard they were stopped by another sentry at a much larger gate flanked by giant ornate statues of centaur-like figures. After they were admitted Philip peered out of the cab's windows as the cab crossed another courtyard, turned right past the vast front façade of a medieval gothic cathedral, continued for another few hundred yards, and stopped in front of a small doorway into a large multi-storied palace. "Here we are at your address, the Lobkovic Palace," announced the driver. Philip happily tipped him and asked Otakar, the driver, for his cell number to call for a ride home. He was glad that he'd arranged to have his phone set up for calls within Europe.

He continued into the main hall where the reception was being held, a large renaissance room with a vaulted ceiling. He recognized Cecily from a distance and walked over to join the group she was with. They were all gazing in awe at the splendid ornamentation of the palace and buzzing about the chamber music recital that was scheduled after dinner.

"Hello, how are you?" she greeted him cheerfully. He was relieved that she didn't seem to be bearing any grudges at his defection from COST. "I just arrived and this reception in a castle is quite an introduction to Prague!" he answered enthusiastically. "I've been to Prague a few times before this on both business and vacation," she told him, "so with all the pubs it's starting to feel a bit like my old British home."

Philip told Cecily that he'd like to catch up with her later and excused himself to find a drink at the bar, exchanging greetings with a few colleagues along the way. In the short queue awaiting the bartender's attention he found himself next to an African-American colleague who inquired affably where he was from. Philip quickly fumbled for the name tag that he had, as usual, stuck in his jacket pocket instead of pinning it on his lapel. "Aren't these tags foolish?" commented the other smiling broadly and introducing himself as Brian Oglethorpe from Ball Aerospace.

Philip was curious what had brought Brian to this conference and he learned that he'd recently been appointed to the Space Science Board and felt that he needed to get up to speed on the highlights of the field. Brian was easy to talk to so they wandered off together to a door open to a broad verandah. The view from the castle hill across Prague was magnificent. Below them, in the dusk they could see the famous medieval

Charles Bridge crossing the Vltava River that divides the city in two. Everywhere were church steeples, crowned by the big baroque dome of St. Nicholas' Cathedral just below them.

"It's hard to get your mind on science when we are surrounded by such splendor, isn't it?" commented Philip. "Yes, it is," responded Brian bemusedly. "Was it Frank Lloyd Wright who said that he would never have produced as much if he'd grown up in Paris, instead of the Midwest where architectural beauty was hard to find?"

"There is some of that in science too," commented Philip. "We are surrounded by so many findings, facilities and opportunities that it can be hard to decide how to proceed. I'm convinced that there are plenty of important problems out there, but it's hard to spot them." They lingered for a few more moments to enjoy the vista before returning inside to join the throng for dinner. It was a seated affair and Philip found himself at a table including a former director of Ondřejov, the Czech national observatory near Prague.

He told Philip that he'd visited the Center for Astrophysics to work with a colleague at the Smithsonian Astrophysical Observatory. "I enjoyed the CFA very much," he told Philip, "I especially liked the beaches of Cape Cod which are better than those on the Baltic or even the Mediterranean!"

Philip smiled sympathetically and asked whether Czech astronomers had the financial support that they needed to travel and for their research. "Oh yes," he was assured, "we do not ask for elaborate facilities, just funds to support ourselves and collaborate on a few international projects."

Philip found this surprising. No director of a US observatory would ever admit that they had all the funds they needed. He wasn't sure whether to find this admirable or depressing. It was a different outlook, he guessed; perhaps it was laudable to admit that you didn't need more money if you were not sure what you needed it for.

Travel fatigue began to get the better of Philip as the dinner wore on and he slipped off to call Otakar. He liked chamber music but it wouldn't do to nod off. He walked out to meet the taxi, gazing upward at the fantastic gargoyles glowering down from the battlements of the Cathedral of St. Vitus. He looked forward to learning more about this city from Ivan and Maika.

Chapter Thirty-Four

Next morning, after a hearty breakfast in a cheery dining room decorated with 19th- century wall paintings of alpine scenes featuring spritely shepherdesses and hearty shepherds, Philip walked over to the metro station and rode to the conference center. He looked forward to hearing the latest results on observations from the NASA Kepler space mission that was monitoring brightness variations of hundreds of thousands of stars, searching for tiny but regular changes caused by planets revolving around them and blocking their light as they transited the star's disk. Those results would be the main focus on the first day of the meeting.

So far, most of the detections had been of planets closer in size to the giant planets of our solar system, Jupiter and Saturn, than to the Earth. These were unlikely to harbor life although their moons might. The big prize would be planets like the Earth in size and also located at a distance from their illuminating star that would produce an environment of the right temperature for life to thrive. Some candidates had been discovered but there were additional criteria on both the planet and star so a larger sample was desirable.

For one, stars much younger than the sun, even those of similar brightness, tended to be magnetically more active. This meant that huge starspots generated by their surface magnetic fields spawned eruptions called flares that ejected beams of energetic particles and high energy ultraviolet and X-ray light that would be harmful to life.

When Philip spotted Cecily in the conference center restaurant at lunch break, he wasn't too surprised to see her seated with Brian Oglethorpe. He'd had a premonition that Brian's recent appointment to the SSB was not random. But he liked what he had seen so far of Brian and was impressed by his sincerity.

Cecily guided them through good choices in Czech cuisine and to show faith, she bravely ordered tripe soup and a glass of Czech white wine that she had enjoyed on previous visits. She told them that this wine was produced by the same Lobkovic family whose palace had hosted their reception last night. Brian and Philip preferred to start slowly and each ordered a goulash soup, a slice of rye bread and a half liter of the famous Urquell beer from Plzen, the original Pilsener.

Cecily asked Brian to lay out for Philip his ideas on how to organize and fund COST. Philip realized as the concept unfolded that Cecily had once again put her diplomatic talents to good use. After Brian had laid out the strategy in much the way he had explained it to Cecily at the SSB meeting, she chimed in and proposed that Philip stay on the project as a Co-Investigator to provide science clout while Brian handled the interactions with Congressional committees that would consume most of the time. "I have all the contacts on file and I'm on first name basis with many of the key characters in the appropriations process. There's no need for you to waste your time on that," added Brian.

Philip told them that he'd think it over and let them know. They all agreed that the lunch had been productive and he and Brian looked forward to further exploration of the excellent Czech beers. Cecily laughed at their gastronomic timidity and recommended that they broaden their choice of Czech foods at dinner.

Privately, Philip had reservations about the project whoever ran it, but Brian and Cecily's strategy had an inevitability about it that he didn't think could be countered. To put it bluntly, always building bigger and newer stuff was the American way and unless Congress and the electorate said "enough" it wasn't within his power to stop it.

Congress had, in fact, said "enough" when costs on the Superconducting Super Collider spiraled out of control and that project had to be terminated. In that case another factor in the cancellation had been Congressional nervousness over sharing US technology with foreign partners. In today's climate that issue could be expected to return with a vengeance.

Philip was a bit annoyed by the suspicion that much of this effort to keep him onboard was motivated by Cecily's desire to keep Harvard University on the COST proposal's cover page. But as he had assured himself many times before, it hardly mattered. He knew how important it had been to his mother that her son was a Harvard professor. Nothing could detract from the closure and peace he had felt telling her that he had achieved tenure a few months before she died.

Philip's mother's family were Jews who had immigrated from Poland around the turn of the century. She had met Philip's father in college and married him soon after, settling in Rhode Island. His family were old line WASPs who were kind to her but never really accepted that their son had failed to marry one of their own. Interestingly, her family felt similarly.

Life at home was reasonably harmonious but Philip grew up never really quite knowing where he fit socially. In high school he realized that their family had been rejected by the country club because his mother was Jewish. He did his best to overcome this stigma by excelling academically and even to some extent in sports. But he had trouble forming enduring friendships and his relations with girls were troubled. Maybe he was trying too hard, his father suggested.

Whatever the reason, Emily was the first woman with whom he was able to form a bond. She wasn't the dark-eyed, pony-tailed ballerina that he'd admired all through high school and had fantasized about marrying. Philip had never mustered the nerve to ask her out. But he and Emily got along well and he had been too busy academically in college to spend time chasing after fantasies.

He looked forward to Emily's arrival but decided that he would make the visit he planned to Prague's famous medieval synagogue and Jewish cemetery before she came. He felt that he needed to do it for his mother so he left the conference center right after the last talk and took a bus to the Old Town.

His visit to the so-called Old-New synagogue and its walled cemetery full of gravestones dating to times immemorial both inspired and depressed him. It was amazing that this community and buildings had survived for a thousand years of ostracism and persecution. But it was depressing because there was little sign that the reasons for the increasingly troubled relations of Jews with Gentile communities in Europe or even in the US could be discussed openly, much less addressed.

He felt that the Holocaust studies that have become part of the curriculum in US schools might be more effective if they gave a balanced account of the history of Jews in Europe, England, and America and dealt more frankly with the reasons for anti-Semitism dating back to medieval times. His maternal grandmother had told him once that for centuries before 1918 Jews had identified with the resented German minority in Central and Eastern Europe, and the Slavic majority never entirely forgave them for that.

Philip was well aware that the main factor in the amazing ability of the US economy to continually re-invent itself was America's five million Jews. Most recently, the tech revolution and the millions of jobs it had spawned was largely the result of the success of Google, Facebook, and many other corporate giants founded by Sergei Brin, Mark Zuckerberg, and other very capable Jewish-American entrepreneurs.

These successes should be celebrated, but Philip knew that they and the issues that caused anti-Semitism could not be discussed frankly because Jews justly feared that broader recognition of their cleverness and influence would invite increased jealousy and retribution. Any minority that dared poke its head too far above the norm faced similar backlash. Successful Chinese and Indian minorities in the US and Asia were equally well aware of the consequences.

Walking back from the cemetery to the metro that evening, Philip stopped for dinner at an unpretentious little place that advertised real Czech specialties. He decided to venture out of his comfort zone and try the Czech standby of roast pork, sauerkraut, and dumplings. It went well with the draft beer from Budejovice, the original Budweiser. He planned to ask why Czech draft beers didn't make him sleepy, unlike the ones back home.

For dessert he enjoyed a small portion of palačinky, the delicate warm crêpes topped with whipped cream and currant preserve. Sitting alone afterward sipping his cappucino surrounded by other diners conversing animatedly in foreign tongues, Philip felt that his mother would have approved of his return to his Central European roots, distant as they were.

Chapter Thirty-Five

The morning of the conference's second day was devoted to highlights of other branches of astronomy, particularly solar research. Looking over the abstracts, it seemed to him that, unlike the concrete advances reported yesterday on exoplanets, the solar talks focused more on the potential of a couple of impressive facilities.

The Parker Solar Probe was to be launched in a few years and was slated to approach to within about four million miles of the sun's visible surface, so well within the outer reaches of the solar corona that is seen at solar eclipse. It certainly was a technological tour-de-force, braving extreme temperatures, and the appeal of sending equipment right into the sun's atmosphere would undeniably grab the popular imagination. But how the probe was expected to solve any important problems of solar research was not explained well in the abstracts.

Still, these days the $1.5 billion price tag barely created a stir and it certainly helped to pay the bills at the Applied Physics Lab at Johns Hopkins University where the probe was being built. Philip had met Eugene Parker of the University of Chicago and he couldn't think of a more deserving scientist after whom to name a NASA mission. Gene had really advanced knowledge of the sun when he predicted and explained the supersonic solar wind in 1958. Unfortunately there seemed little prospect that his namesake project would do as well.

The other major solar project was a giant ground-based telescope named not after a respected scientist, but after the Congressman who helped push it through the budget process and past the throngs of native Hawaiians who protested its construction on the sacred mountain Hale-akala. Here again, the promise was that its larger aperture would help understand the mysteries of the sun. The details were somewhat vague but much of its $344 million cost was a windfall from an Obama-era economic revival set-aside, so it would have been un-American not to build it.

"Sic transit gloria scientiae," thought Philip as he absorbed these abstracts over breakfast. In past years he had been impressed with the advances in understanding of the solar interior through the development of helioseismology, of the sun's luminosity variation through radiometric monitoring, and of the huge bubbles of ionized gas called solar transients, observed being expelled from the sun using clever coronagraphs. "I guess they just couldn't think of any more lacunae in our understanding of the sun," he concluded.

Philip didn't mind the walk to the subway. He hoped that the lovely spring weather would hold for Emily's visit and enjoyed people-watching on the metro. Observing the happy antics of Czech youngsters in their Red Sox hats and latest model Nike sneakers he wondered whether they gave any thought to the grim Communist dictatorship that had ruled this now sunny land when their parents were their age.

One of the Czech astronomers had told him that, if you look at Czech history, you are motivated to "have dessert first." "That is, if you are living in this country where seldom more than a few decades have gone by between catastrophes, you are skeptical of long-range planning for the future," he explained. "Either the Germans or the Russians will invade; it is inevitable. That is different from the USA where you have had relative stability for centuries. Then it makes sense to build for future generations. Here, it makes more sense to grab what you can while the grabbing is good. Persuading Czechs to invest in their own country is still difficult."

Philip had decided to pass up the bus excursion to the Karlštejn royal castle planned for that afternoon. He needed to work on his talk for tomorrow morning. Also, he wanted to poke his head into the little museum of Alfons Mucha's art that he had seen advertised in the tourist literature handed out to the conference attendees. He wasn't sure whether Ivan and Maika would consider that a worthwhile destination but he'd always liked Mucha's posters that hung everywhere in college dorms when he was a student. He'd join his colleagues again at the conference banquet at night.

Philip enjoyed walking through downtown Prague and on his way to the museum, he stopped for lunch at the Obecny Dum, the grand Art Nouveau Municipal Building that housed a large theatre and a beer hall style restaurant in the basement that the receptionist at Villa Lanna had recommended for Czech specialties. Before descending he looked upward to admire the famous ceiling paintings by Mucha, reminiscent of John Singer Sargent's well-known ceiling decorations in the Boston Public Library.

The specials that day included a soup of chicken liver dumplings in a clear broth for starters followed by svičkova, the Czech version of sauerbraten served with dumplings and red sauerkraut. He was pleased at having tried both, washed down with some excellent local Smichov beer. For dessert the waiter recommended the house strudel with real whipped cream, and Philip was not disappointed.

He knew a little about Mucha's life in Paris, how he made his mark with his famous poster of the American actress Sarah Bernhardt and subsequently with the celebrated murals he painted to decorate the new Paris Opera House. All this and more was well covered in the Museum, he discovered after finally gaining admission. But he was most drawn to the pictures of the artist's family life in the years when he lived with his family at the rented Castle Zbiroh near Prague and produced his masterpiece, the twenty huge panels describing the history of the Slavs. A series of photographs depicted the production of this Slavonic Epopée, as it is named. Mucha used live models in this work, including his wife, son, and teen-aged daughter.

Philip noted in particular a couple of easily missed pictures hanging in a corner showing Mucha standing behind his daughter, cradling her partially clad torso with his hands placed just below her breasts. Not the sort of thing that you'd want to display these days, he thought to himself. The text described how Jaroslava, his daughter, worked closely with him mixing paints and modeling. There was no mention of what she did later in life.

This struck Philip as odd since the life of her younger brother, Jiri, was described in some detail. Philip had read elsewhere that Jiri had indeed led an interesting life with intimations that he'd been a double agent working for both the Czech and British Secret services. This had only recently been discovered from material made public in the opening of the Czech secret police files and was not mentioned in the museum's material.

Philip asked one of the museum staff about Jaroslava but was told that only the curator could answer such questions and she was not available. So he left the museum pondering this strange lacuna in Mucha family history. Still pondering, he walked to the subway station and took the metro back to the Villa Lanna. But he spent the rest of the afternoon working on his talk and quite forgot about Mucha's daughter.

Chapter Thirty-Six

That evening Philip once again called Otakar and gave him his destination for the conference banquet. It was in the Vladislavsky Sal; the 15th-century main ceremonial hall of the Hradčany Castle. They followed much the same route up the Castle Hill and past the sentries but stopped in the main courtyard before a small entrance into the Presidential Palace. Inside, the guests were wandering around the vast gothic hall, admiring its 50-foot high vaulted ceiling and elaborate renaissance windows.

He was seated for dinner at a table that included a Czech theoretical physicist who had spent a year at Caltech back in the 1970s. Philip wanted to know whether it had been difficult for him to return to Charles University. "I enjoyed working with the Caltech physicists, of course," he responded. "But you have to understand that many of the ideas that they were working on actually originated in visits to Moscow to work with brilliant Russians like Novikov, Zeldovich and Shklovsky. The originality came largely from abroad, less from Caltech."

Philip wasn't aware of this but Sandra Wells added that she wasn't surprised. "We are victims of our own self-referencing in other areas too," she told them. "When we started offering our Stanford University physics course on the Internet, the first time we gave an exam we were stunned to find that dozens of foreign students did better on it than our own highest scoring Stanford undergraduate. These were kids from Pakistan, from Romania, from all over the planet. There is a tremendous amount of talent out there, outside the range of our highly touted US schools," she told them.

Nigel Worthington wondered about the proliferation of large projects. "Would Ernest Rutherford have chosen to pursue a career in physics if it had meant working at CERN on the Large Hadron Collider with 10,000 others?" he mused. "The danger that the increasing size of projects is robbing us of the most talented minds is not being faced squarely," he continued.

"The younger generation may not agree with you," suggested Sandra. "Many of them seem to enjoy working in the anonymity of large groups and making creative use of mass communication via the Internet. Some of them spend much of their time involving the public by financing research with crowd funding and focusing on social relevance, inclusiveness and diversity rather than elitism and selectivity.

"It may be that by including the hundreds or thousands of students—many in developing countries, some of whom surpass the performance of our best at Stanford—we'll end up with as much excellence and as many advances in understanding the cosmos as we ever did in the past."

A Chinese astronomer at their table who had so far been entirely preoccupied with the excellent roast duck, commented that he had done his graduate studies at Berkeley. "We have doubled the Chinese budget for science over a decade but the yield of world class results has been disappointing. Whatever the failings of US science, be careful to appreciate its achievements." Philip wasn't sure that he was so optimistic, but it was time to call it a night. He needed to be fresh for his talk tomorrow, the first of the morning session.

Chapter Thirty-Seven

The cosmologists on the conference's Scientific Organizing Committee had invited Philip to give an overview that would make clear to the attendees from all fields of astronomy why cosmology was still where the most exciting advances were being made. Philip wasn't sure that this was the case but he didn't want to let his colleagues down.

Assembling his material he found to his relief that he could use much of what he had already prepared for his undergraduate course at Harvard. Outside their narrow specialty, his colleagues knew little of happenings in other areas. He himself knew little of galactic, much less planetary or solar astronomy. His faculty partner in teaching the course handled that material. It was even a point of pride amongst cosmologists to show off their ignorance about what they considered the less elite aspects of astronomy.

His talk would not tell other cosmologists anything they didn't already know but his mandate was to inform the others. Philip was past needing to impress audiences with his cleverness, which made it more likely that they might actually learn something. As the room filled he was gratified that so many of his colleagues, including some who had disagreed with him in the past, had made the effort to get to the opening talk rather than dawdling over breakfast.

Philip began the presentation by briefly summarizing the status of cosmology a few years ago and then focused on the recent highlights. These included advances in understanding supernovae that were used to establish the distances of very distant galaxies, enabling more accurate assessment of evidence for the mysterious acceleration of the universe's expansion. Other work provided improved limits on neutrinos as a possible source of the missing mass that holds galaxies and the universe together. Detection of gravitational waves by the LIGO interferometer demonstrated yet further that General Relativity gave a good description of gravity, useful to know as one ventures into the gravity-dominated present universe.

Looking out at the audience Philip saw few of them gazing into their laptops, which suggested that his talk was holding his colleagues' attention. There were a few polite questions afterward but the audience seemed satisfied. When he was younger he used to feel a flood of relief after an important talk and the subsequent high lasted for a few hours. For better or worse these days life just went on.

Philip slipped out after lunch. Emily was arriving next morning and he wanted to buy her some flowers to greet her on arrival. She had sounded depressed when he called her the day before. He didn't know whether it was related to her elderly parents' health or some issue at work. She seemed to need cheering up.

He'd been advised of a flower shop on the Vaclavsky Naměsti—Prague's Times Square or perhaps better, Trafalgar Square. It wasn't considered a particularly pretty part of town and it was best to keep your hand on your wallet, but there was often something interesting like a political rally or rock concert going on so it was worth a visit.

When he emerged from the flower shop holding six red roses, he decided to walk down to the Vltava and cross the medieval Charles Bridge embellished with life-sized statues of saints. It began to drizzle and he was tempted to just take the metro straight back to the Villa Lanna from the Mustek station at the bottom of the Vaclavske Naměsti. But he decided to walk on and was rewarded.

As he emerged from the narrow Old Town street leading to the bridge he was greeted by a splendid sight. Beams of sunlight shone through the shower and a grand rainbow arched over the towers of the bridge and the imposing bulk of the Hradčany Castle on the hill just beyond. Philip joined the thousands of tourists packed in around him gazing in wonder at this breathtaking sight.

He wished that Emily were here to witness this. At least she'll have the flowers, he thought, as he continued on his way across the bridge and then turned right along Valdštejnska Street towards the metro home. By now the roses were looking a bit tired but he got a small vase from the receptionist and put them on the windowsill in his room to recover.

Chapter Thirty-Eight

Philip had offered to meet Emily on arrival at Prague's Havel Airport but she maintained that she was quite able to take a cab on her own and please not to incur the extra cab fare. So he contented himself in arranging for trustworthy Otakar to pick her up and bring her to the Villa Lanna. He told her that, since her flight was coming in around noon, he would order lunch at the hotel so she could rest afterwards.

Maika had asked them whether they would like to see a performance of the well-known Smetana opera, *The Bartered Bride*, at the National Theatre. It was a favorite event for both Czechs and tourists. Unfortunately, the only performance during their stay was scheduled for the evening of Emily's arrival, but perhaps she could manage if she took a nap in the afternoon.

Emily arrived safely and was delighted with the Villa Lanna and their room. Since the weather was cooperative, they had lunch served on the broad verandah and enjoyed the spring sunshine. Philip was encouraged by Emily's more cheerful mood.

They met the Vesely's for early dinner at the fin-de-siécle Slavia Café across the street from the theatre and Emily was charmed by its décor. She quizzed Ivan on his connection to the Villa Lanna and was told that his great-grandfather had done well in the oil business around the turn of the century and had bought the architecturally significant villa in 1915 as a hedge against the inflation that he expected after the war ended. He also bought a working farm outside Prague to provide food to the family and servants when the shortages that he also foresaw occurred during WWI.

After the war the villa was rented to the French Army for the use of General Pellé who was entrusted with the training of the army of the newly constituted Czechoslovak Republic. It was eventually sold by the family in 1927. "It is now used by the National Academy to house foreign dignitaries like you and Philip," said Ivan with a smile.

The rollicking opera played to a packed house and during the intermission Emily enjoyed the varieties of colorful national dress that several members of the audience had worn to the performance. Maika explained that the various regions of the country each had their own design of the elaborately stitched full skirt and pleated blouse. "I have one at home but I don't fit into it as easily anymore as I did when I was younger," she told Emily, laughing.

Ivan told them after the performance that he and Maika were happy to be at their disposal tomorrow to view whatever Prague sights they wanted to take in. "But if you prefer to have us make some suggestions, I have some favorite places that I think you would enjoy." Philip and Emily said that they would be delighted to follow his suggested itinerary and they agreed that Ivan and Maika would stop by at the Lanna tomorrow after breakfast to pick them up.

When the Vesely's arrived next morning, they joined Philip and Emily for coffee in the dining room. Sunlight was streaming through the tall windows and it looked like another bright day. Ivan suggested that they start with a visit to St. Vitus' Cathedral in the Hradčany Castle, which Emily had not yet seen. "It is the highest priority of any tourist to see the Castle and we should get up there before it gets too crowded," he told them. Philip said that Otakar was available to pick them up whenever they were ready.

Emily was as amazed by the renaissance palaces in the Castle Square and the Castle itself, as Philip had been, and they marveled at the solemnity of the gothic cathedral. Ivan explained that it was one of the last to be undertaken, over a century later than Chartres and Reims in France. It was one of the projects of Charles IV, the beloved Czech king who became the Holy Roman Emperor in the mid-14th century. "He also founded the Charles University, the world's oldest after Oxford and

Bologna, and built the famous bridge across the Vltava that we will cross later," he added.

As they exited the Castle compound Ivan told them that the palaces that surrounded the expansive courtyard before the gates had belonged to the most prominent Bohemian families such as the Schwarzenbergs and Cernin's, and the Lobkovic family's palace where the conference opening reception had been held, was within the compound of the Royal Castle itself.

"Many people were surprised that the government was so generous in returning the estates of the former aristocracy," Ivan said. "They consider the nobility a useless relic of the distant past. But there was a situation not too long ago when they came in handy. Near the end of WWII about half a million Wehrmacht troops were retreating from the Eastern Front through Prague. They desperately wanted to surrender before the pursuing Red Army caught them, but not to any old Czech street fighter.

"Fortunately one of the Czech partisans was the young Prince František Schwarzenberg whose ancestor, Prince Karl Philippe Schwarzenberg, Grand Marshal of the Austrian, Prussian and Russian Armies, had defeated Napoleon Bonaparte at Leipzig in 1813. When the Major in charge of the strategic Prague railway station offered to surrender it to an officer of at least equivalent rank, the Czechs brought forth the Prince. As expected, the SS Major was more than willing to surrender to Prince Franz von und zu Schwarzenberg."

Ivan wanted them to see the lovely library of the Strahov Monastery located on the ridge not far from the Castle. As they strolled in that direction he gave them a little background on Czech history. "The Czech lands, as they have been referred to since the Middle Ages," he began, "consisted of Bohemia, where we are now, and Moravia, a slightly smaller region to the East. At various periods parts of Silesia, located to the north, were also included.

"These Czech lands," he continued, "became a kingdom that remained one of the wealthiest and most influential of Central Europe throughout the Middle Ages. The discovery of some of Europe's richest silver mines at Kutna Hora near Prague in the 13th century guaranteed the prosperity of the land. But it also attracted the attention of acquisitive neighbors like the Austrians and Hungarians.

"Also the king encouraged immigration of a large population of Germans to help work these silver and also iron and coal mines," he went on. "The Germans tended to be more enterprising and they gradually took over the upper social levels of Prague and even the smaller towns. This created animosity between them and the Czech population.

"In the early 15th century the Czechs became dissatisfied with the corruption of the Catholic Church," Ivan explained, "and when their intellectual leader, Jan Hus, was burnt at the stake in 1415, the Hussite movement took over the land. They destroyed churches and monasteries and set up the first Protestant state a century before Luther. Their movement was also directed towards reducing the German influence.

"Led by their talented military leader Jan Žižka," Ivan continued, "they held off several Crusades sent against them by the Pope and raided into neighboring Austria, Hungary, Poland, and Bavaria. As often happens though, they eventually turned on one another and their power declined towards the end of the 15th century. But the more moderate Protestants managed to maintain a strong influence on Czech government until 1620."

As they were approaching the Strahov Monastery, Philip remembered the enigmatic parting comment about Prague architecture made by his neighbor on the flight from Munich a few days earlier and made a note to himself to ask Ivan later what that might have meant.

Chapter Thirty-Nine

After viewing the magnificent baroque ceilings of the Strahov libraries, they descended the steep hill towards the river. Next Ivan wanted to show them his favorite church, the St. Nicholas Cathedral. "It is special to me since it is where my parents were married," he told Emily and Philip as they came to the square bordered from above by the Liechtenstein Palace and below by the vast bulk of the baroque cathedral. "It is nothing special from the outside although the dome dominates the whole neighborhood, but wait until you see the inside," he told them.

It was, as promised, a breathtaking baroque interior. Imposing in its size but especially in the ornamentation; ornate without passing over into vulgar. "The stone used here was warmer in color than the cold grey of some other otherwise splendid baroque cathedrals, like the famous Dom in Salzburg," Ivan commented.

Ivan and Maika walked slowly to the front and sat in silence for a while in the front pew. "I like to remember my parents who were married here in 1944. It was a cruel year, the time of the Heydrich assassination by Czech partisans, and the savage reprisals of the Nazis," explained Ivan.

Philip thought this might be a good time to ask Ivan his question about Protestants but Ivan suggested that the answer wait until they sat down to lunch. He and Maika guided them to a pleasant outdoor café restaurant nearby in the little Square of the Knights of Malta.

After they found a table in the sun and rearranged the chairs to optimize their exposure, Ivan recommended the house Krušovicky Velvet beer and suggested the chicken liver dumpling soup. "It is called leberknödel suppe in Germany and Austria," he informed Emily, thinking that this might reassure her. "A nice bowl of that with some assorted open-faced sandwiches and cucumber-dill salad, would make a good light Prague lunch for you," he recommended.

After they ordered, Ivan turned to Philip's question. "I mentioned that the Protestant influence here continued until 1620," he began. "In 1618 some of the Protestant nobles got upset over the failure of the Holy Roman Emperor to continue granting them basic rights of worship and they threw two of his top councilors out of one of the windows in Hradčany Castle. That was the so-called Defenestration of Prague which set off the Thirty Years' War between Protestants and Catholics that caused great devastation of Central Europe until the peace of Westphalia in 1648. It was the most destructive war the Western world had seen to that day.

"The key battle as far as the Czechs were concerned was fought in 1620 at the White Mountain, a hill outside Prague," Ivan continued, heartily attacking his main course of svičkova and dumplings. "The Protestant side lost to the Catholic forces of the Hapsburg Emperor and the Pope. After that Protestantism was suppressed in the Czech lands and tens of thousands of the country's elite left. Amongst them were the sect later to be known as the Moravian Brethren who emigrated first to Germany and later continued to America where they founded cities like Winston-Salem, North Carolina.

"A particularly notable emigrant was Jan Amos Komensky or Comenius, the most famous humanist educator of his time," he said. "John Winthrop offered him the first Presidency of the recently founded Harvard College, but he preferred the more attractive offer of the King of Sweden to teach at Uppsala University.

"The Czech nobility, which was largely Protestant, lost their lands and most of them emigrated," he added. "The Catholic nobles who had remained loyal to the Emperor, like the Lobkovic family, amassed huge wealth in taking over these confiscated properties. Foreign mercenaries who had served in the Hapsburg army like the Italian Colloredo brothers and the German Schwarzenbergs, were rewarded with imposing titles and huge estates and became some of the most powerful aristocrats of the land.

"What your friend was probably alluding to," Ivan continued, "was the effort that the Jesuits made to impress Czechs with the majesty of Catholicism by building beautiful structures like St. Nicholas' Cathedral. All the splendid baroque architecture that you see here was built during the Counter-Reformation in the 17th and 18th centuries to wipe out any vestige of nostalgia that Czechs might have harbored for their Protestant religion.

"So," Ivan added, "to this day, there are devout Protestants amongst my relatives who refuse to acknowledge the merit of baroque churches found throughout the Czech lands. They see in them the oppression of their religion in the three centuries of the Hapsburg domination of this country, until the end of WWI in 1918. You might find it interesting to meet one of them, if I can arrange dinner with him tonight. But let's move on now across the Charles Bridge, to visit the Old Town Square and maybe have some coffee and dessert along the way."

As they navigated through back streets from the Maltese Square to the bridge, Ivan filled them in on its history, dating back the beginning of its construction in 1357. "When Prague was hit by the disastrous flood of the Vltava in 2002, there was concern that this and the other bridges might not hold, but the seven centuries–old abutments held just fine."

"Towards the end of the Thirty Years' War the Swedish Army penetrated as far as Prague and captured the Castle," Ivan told them. "But their passage across the Vltava to the Old Town was prevented by troops led by Field Marshal Rudolf Colloredo, who went on to power and riches as a Bohemian grandee," Ivan continued. "The family later joined by marriage with the wealthy German Catholic family, the Mannsfelds." Philip and Emily were interested to hear that some of the Colloredo-Mannsfeld family (now spelled Mansfeld) have since become prominent in Boston.

After crossing the bridge they wended their way to the Old Town Square and stood with hundreds of others waiting to see the famous medieval clock Orloj perform its hourly pantomime of carved wooden figures. Afterwards, Ivan suggested that they stop at a café on the square for some coffee and people-watching. He also called his cousin, Dalibor, to finalize plans to meet for dinner that evening.

Instead of pastry Ivan ordered a serving of fruit dumplings, a Czech specialty. Each "ovocny knedlik" as they are called, contained an apricot, strawberry or plum and seeing what a hit they were with Philip, Maika promised to give Emily the recipe. While they were savoring this nice afternoon break Philip remembered to tell them of his visit to the Mucha museum and his unsuccessful attempt to find out about Mucha's daughter.

"That's quite a coincidence," Ivan laughed, "It happens that the Bugatti car I need to look at is located at Zbiroh Castle, which is where Mucha lived for many years with his family while he was painting the Slavonic Epopée. I was planning to drive us up there tomorrow. It's not far out of Prague, just off the main highway towards Marienbad. There is an interesting story about Mucha that I'll tell you on the way up; it also involves his daughter."

Chapter Forty

They met for dinner that evening at the Obecny Dum where Philip had enjoyed lunch during the conference. He wanted to show Emily the fin-de-siècle architecture and the fine ceiling paintings by Mucha. This time they tried the more elegant restaurant on the ground floor rather than the beer hall in the basement where Philip had eaten.

Ivan suggested that they have a seat in the bar because his cousin Dalibor and wife Zdenka would be a bit late. They ordered some of the white wine from the Lobkovic vineyards in Mělnik, near Prague and Ivan filled them in on Dalibor's background.

"Dalibor's father was my mother's brother," Ivan explained. "In 1948, after the Communist putsch, when my parents emigrated to Canada, Dalibor's parents were politically more to the left and they believed in a Communist Czechoslovakia, so they stayed. They were devout Protestants and had a strong social conscience. His father became a doctor and to atone for his capitalist upbringing the family spent their early years in a particularly grim coal mining area near the East German border. Over the years Dalibor's views have evolved but I wanted you to hear his view of Czech 20th-century history from the point of someone who lived here for the past decades, not an émigré like me."

At this, he spotted Dalibor and his wife Zdenka entering the restaurant and rose to greet them and introduce them to Emily and Philip. Dalibor was a tall, lanky, and slightly hunched elderly man who greeted them with a friendly smile. His younger-looking wife was also welcoming. She spoke remarkably good English and his was also passable. Ivan had told them earlier that their English proficiency was in part why he wanted Philip and Emily to meet them rather than some of his numerous other relatives whose English was weaker.

Ivan had managed to get a table in a quiet nook and Dalibor requested, smiling at Philip, to sit so his better ear was towards the American visitors. They ordered more wine and looked over the menu. "The difficulty in the better Prague restaurants," Maika commented, "is that they consider French or Italian dishes more upscale. I wish they would leave boeuf bourguignon to the French and serve the Czech cuisine that customers like us would prefer!" Nevertheless, they were all able to find dishes to their liking and Ivan asked Dalibor to tell Philip and Emily what work he and Zdenka were doing and how they had arrived at their choice of career.

Looking to his wife from time to time to help with the English, Dalibor told the visitors how happy they were to meet them and hoped that their stay in Prague had been pleasant. "I hope also that my energetic cousin has not been tiring you out," he said, smiling at Ivan. "It is important in Prague to take time to just walk around the city with no specific itinerary, just go where your curiosity takes you."

"If you are really as interested as Ivan suggests," he began modestly, "I am a minister of the Evangelical Church, which is what you would call Protestant. Zdenka is a doctor. We both come from families in which religion has always been important. My grandfather-in-law was an influential theologian known world-wide for his belief that Christianity could be reconciled with Communism. He spent considerable time in the US and even taught at the Princeton Theological Seminary during the WWII. They travelled widely after the war; that's why Zdenka speaks English so much better than I."

"We are no longer as confident as my grandfather-in-law that the Christianity and the Communism could be reconciled," he continued. "After Soviet invasion of Czechoslovakia in 1968 he also had change of heart. But we still believe both in dignity of all mankind and seek to achieve the harmony between the Christianity and egalitarian socialist state."

Ivan remembered the interesting conversation that he and Philip had had with the technician Abe Reeves at the Mt. Wilson Observatory.

"Philip and I have become intrigued with a different connection—the relation between science and faith," he said. "Is that something that theologians like you have considered?"

"Certainly," answered Dalibor. "It is frequent topic at our meetings. But what exactly is context in which you are interested? I suspect that our concern, that religion is being replaced by the science, is not what you have in your mind."

"Right," said Philip, "we are trying to understand whether a preference for faith in a Divine Creator over the multiverse explanation of the universe's beginning, for instance, might be caused by a fundamental difference in brain wiring. It might instead be ascribed to something more banal, like conservatives' dislike of liberal scientists."

"It seems to me that both probably contribute," responded Dalibor. "We certainly see plenty of the more banal explanation, as you call it, around us here in Europe. But it is interesting that there are liberals here and in America, who are also religious. And we see many conservatives who have no use for the religion."

"An explanation must cover the intermediate cases, as you say," Ivan commented, "but it is important not to forget the startling statistic that, in the US, about forty percent of the population prefer Creationism over Darwinian evolution. That's almost one hundred forty million people."

"I agree that it is highly interesting phenomenon that deserves more attention than it has received," Dalibor said. "In the US," said Ivan, "it is more than an academic question. It seems to be a significant component of the general disaffection between conservatives and liberals that is making the country increasingly difficult to govern."

"Preference for what is called the populism in much of Europe," Dalibor said, "is caused greatly by immigration issues but is deeper. Schism between Protestants and the Catholics that destroyed the center of Europe already in the 17th century maybe also could be ascribed to difference in brain wiring, as you call it.

"Protestants were mainly the Northern Europeans who believed in relatively ascetic, I think is the word, way of life. Their churches have not much ornamentation. Catholic mentality of France, Bavaria, Italy, and Spain preferred more rich tastes, like the baroque. Northern Protestant tendency to defer gratification, as you call it, was not so much popular in south.

"Divisions we see now are made by a bad feeling of the part of the population who believe in the deferred gratification toward the people who prefer to have the dessert first, also to borrow an expression I have learned from American television." Dalibor said. "How the preferences for faith over science might be connected to that is a complicated question."

"Unfortunately, it is one thing to diagnose these differences but another to do something about reconciliation," Ivan interjected. "The only remedy that I've seen to work is to show good will by concrete actions rather than just words. If scientists showed that they are willing to follow scientific method even when the results make them uncomfortable, conservatives would be more likely to sit up and take notice."

The staff seemed to be hoping to close early, so Ivan looked at his watch and proclaimed that they should probably allow their American visitors to get some sleep. "We have a busy day tomorrow," he reminded them, "driving out to Zbiroh Castle to have a look at this Bugatti!"

Dalibor was surprised that his American cousin was collecting such fancy cars. "Philip and Emily," he said, turning to them, "you are fortunate to have Ivan and Maika to show you our country. But remember that there are the Czechs like Zdenka and me who have been able to raise a big and happy family in this country with no car at all!"

Chapter Forty-One

Next morning Ivan and Maika picked up the Frobishers after breakfast and headed west out of Prague. Ivan explained his excitement at seeing this Bugatti. "When I was a kid," he began, "I was immersed in the lore of these beautiful race cars and grand touring sedans of the 1920s and '30s. My father had fallen under their spell as a child in Moravia."

"He recounted to me the exploits of the legendary Elisabeth Junek, the foremost woman racer of her time. In her greatest triumph she nearly won the grueling Targa Florio road race in Sicily in 1928. She was leading the field of the most famous male drivers when she collided with a rock that had mysteriously been placed in her path and then removed before the other racers arrived. Her mechanic, travelling with her, managed to repair the damage but she lost her lead and finished fifth.

"When my father turned seventy," Ivan continued, "I arranged to have him invited to a meet of Bugatti cars at the Rockefellers' summer property in Camden, Maine. I had heard that the American Bugatti Club had invited the eighty-ish Madame Junek as an honorary visitor from Prague. She wasn't proficient in English so my father became her official interpreter. It was a memorable experience for all of us.

"So you can imagine my excitement when I heard that the car that she had driven in the Targa Florio had popped up in a barn of this castle. Apparently it is in good condition. I hope that we will not be disappointed."

It was not a long drive and although the weather had turned drizzly the sight of the expansive castle and grounds was impressive. "There used to be a fortress here already in the Middle Ages," Ivan told them, "but as often happened it was rebuilt centuries later to suit renaissance tastes, so it's now really more a grand palace than a castle.

"It is operated now as a hotel but it was purchased in the late nineteenth century by the Colloredo-Mansfelds whom I mentioned yesterday. As I think I also mentioned, the painter Mucha rented part of the castle from them in the teens and twenties of the twentieth century because he needed a lot of space to create his giant twenty-panel masterpiece, the Slavonic Epopée."

Ivan had arranged to meet the car's seller in the hotel's reception area and after introductions they walked through the grounds to a small outbuilding. When the garage door was opened Emily exclaimed that the car, gleaming in Bugatti's signature light blue and polished chrome, was the most beautiful piece of machinery she had ever seen.

It was smaller than Philip had expected, but Ivan explained that its light weight and powerful engine made it both fast and nimble. Maika appreciated how no detail had been overlooked. When the hood was lifted the engine was as beautifully fabricated and polished as the exterior.

Ivan asked, somewhat diffidently, whether it was in driving condition at this point. The seller said it was and he'd be happy to take him for a short spin, but he couldn't let Ivan drive. "The insurance does not allow that," he apologized. They felt awkward asking for a ride when it was rainy and muddy, but the seller suggested that they have lunch at the hotel and they could return later when the weather was expected to improve a bit. He had some errands to run and would return to meet them at the hotel.

At lunch, Emily asked Ivan to tell them the story he had mentioned yesterday about Mucha. Ivan preferred not to go into all the details but said that Mucha's daughter, Jaroslava, claimed that Adolf Hitler was her father's illegitimate son.

"Her brother, Jiri, seems to have tried to have her declared insane and institutionalized and he gave alternative explanations for her evidence. But he himself was a colorful character to say the least and may not be the most credible authority."

"Wow! That is a strange story," exclaimed Emily. "Have you tried to check it in any way?" "I tried learning more about her from the Mucha

Foundation that runs the museum but as Philip also found, they were not helpful. Personally I wonder from what Philip saw in the photographs of Alfons and her displayed in the museum whether their relationship was entirely healthy. If not, she may have grown to resent her father and manufactured this story to harm him. It wouldn't be the first time that a famous person's life included some regrettable episodes. Why else would she make up something as strange as this?"

Philip commented that strange tales of unexpected paternity seem to be sprouting in the Czech Republic. "I just read in the Prague English-language newspaper that T. G. Masaryk, the venerated founder of the Czechoslovak nation in 1918, may have been the illegitimate son of the Austrian Emperor Franz Josef! If true, that could put a new spin on 20th century Czech history."

As they were finishing their coffee Maika waved to the seller of the Bugatti who had just entered the dining room looking for them. "The weather is better, so let's go for a little ride," he offered cheerfully.

When they walked back to the barn he opened the large gate and held the car door for Ivan. It was a tight fit even into the passenger's seat and Ivan, who was considerably bulkier than the seller, began to doubt that he could actually drive this car. Nevertheless, he waited expectantly as the owner started the engine. It took a while to catch but the sound was exhilarating even at idle. When he clambered in and put the engine in gear it rose first to a loud clatter and then as they shot off down the road, to an impressive shriek. Maika laughed as Ivan's visored cap sailed off as soon as the motorists got going.

They soon disappeared from sight down the curved castle drive, but could be heard for quite a while longer. They returned in a quarter of an hour a bit dustier but from Ivan's smile Maika saw that he was hooked. "It's a wonderful machine," he told the seller in Czech. "I'm too big to drive it myself, but my wife is an excellent driver; she could zip me around like Mme. Junek!"

At this surprising turn of events they thanked the owner, took a few photos of the car and headed back to Prague. It had been a full day and they said goodbye at the Lanna. The Frobishers had booked an overnight excursion on the next day to Česky Krumlov, the dramatic medieval town in the mountains near the Austrian border. Ivan told Philip that he'd get in touch after they were all back home. "I've done some thinking about what I hope we could get involved in together," he said, "and I'd like to discuss it with you soon."

Chapter Forty-Two

For a couple of months after Anna arrived in St. Petersburg she was happy just to vegetate. She had worked hard at Harvard and mainly, was still recovering from the trauma of her internship in Washington. She wasn't able to dispel the suspicion that Cecily had used her and wondered also whether Philip had been in on the plot. He didn't seem to be capable of such a scheme, but who knows, Cecily might have pulled him into it.

At first she still felt some obligation to NASA and visited the Pulkovo Observatory once or twice a week. But she wasn't able to find anything worthwhile to do there so she gradually spent more of her time exploring the museums and sights of her native city.

She discovered the wonderful portrait of the Soviet-era poetess Anna Akhmatova in the Museum of Russian Art and re-visited it several times, thinking that she saw a resemblance to herself. She also discovered the side entrance to the Hermitage Palace where she could get in without waiting in the interminable queues of tourists. She remembered visiting the Palace with a school group as a child, but she had been too young to appreciate its splendor and also its decadence.

The city's vast expanse and the bridges over the Neva river were amazing to her under all illuminations from bright morning sun to the brief darkness of the far-northerly city's night when the lights shone out and the river flowed inky black between the illuminated banks. The battleship *Potemkin* anchored at the quay reminded her of the ferment of the Bolshevik revolution and stirred her youthful imagination.

Anna's grandparents' apartment was small but she had her own room and she enjoyed talking with them about things that her parents seemed reluctant to discuss. Why had they emigrated? Did their parents approve and could they have left with them?

She found the answers disconcerting. A friend who had managed to get out in the 1990s had apparently told them that they would make more money and have a better life if they took advantage of the opportunity to emigrate. Now, her grandparents told Anna, they do make more money but it is less clear that they are happier. "Your mother writes that they have trouble making friends with Americans," her grandfather told her. Anna had to agree because after over a decade in the USA they still associated almost entirely with other Russian Jewish emigrés.

When she asked whether they had been victims of anti-Semitic persecution her grandmother told her that some of the emigrés probably had been but she wasn't aware that her parents had been targeted. "It was a convenient chance to get out and live a more comfortable life," her grandmother told her. "We had no interest in going along. All the rest of our family and friends are here and we are quite happy in Russia."

As spring turned to summer Anna found herself wanting to get involved in the young life of the lively city. She found a part time job doing research for Google and met some people of her age. One evening a new friend from the office suggested that she come along with a group to explore some of the hot new dance clubs. She had a great time and met a good-looking, well-dressed young man named Igor. "I have been watching you since you entered," he said, leaning over towards her conspiratorially at the bar, "and I think you are the prettiest girl here." Anna laughed at this obvious pick-up line but she liked his smile and broad shoulders and accepted when he asked her to dance.

Afterwards Igor bought her a drink and asked somewhat disinterestedly, she thought, about her background. He took in stride her revelation that she was actually an American although born in St. Petersburg. He only nodded when she confided that she was living with her grandparents for a few months. She expected him to ask why she had come back here if she had a degree from Harvard University, but he didn't.

Anna refused his offer to take her home, saying that she had promised her girlfriend that she would return with her. But she accepted Igor's invitation to dinner the next day. As she left the bar she realized that Igor hadn't told her anything about himself. After four years around Harvard undergrads who hardly ever stopped marketing themselves, she found that refreshing but a bit odd.

She was on her guard when they met in the park but it actually felt quite natural to have Igor put his arm around her waist and guide her in the direction of the restaurant that he had chosen. It was a lovely warm summer evening and at the city's northerly latitude the sun was still high. The light green poplar leaves shimmered in the warm breeze and Anna felt happy and in a strange way, satisfyingly Russian. Igor asked her whether she liked Italian food; a nice new trattoria had opened recently and he thought that they might try it.

At dinner she made a point of asking about him. His answers seemed a bit vague; he worked in research for the government. He said that he was originally from Moscow but had been transferred to St. Petersburg two years ago. "So we are both recent arrivals here," he told her smiling broadly. They talked about shared interests in sports. He was an avid skier and she had just started.

Anna limited herself to two small glasses of wine and turned down Igor's invitation for a nightcap at his apartment. She told him with a friendly smile that she hoped to get to know him better and suggested that they visit the Winter Palace together over the weekend. She was confident that, after gazing longingly into her green eyes over the excellent rigatoni carbonara all evening, he was unlikely to wander off on her.

Chapter Forty-Three

When Ivan had returned to Cambridge in the spring his mind was made up on what he wanted to do as a project with Philip and the Center for Astrophysics. He had discussed it at length with Maika in Prague after the dinner with Dalibor and Zdenka, and then afterwards on the flight back to the US.

He and Maika had stayed on for a few more days in Prague visiting relatives and arranging to have the Bugatti shipped to the States. Ivan looked forward to Maika driving it in the annual hill climb in Wellesley west of Boston next May. He'd be happy to video a pretty woman racing a lovely car. A fitting continuation to a long tradition, he thought to himself.

Evenings they strolled along the river on the Mala Strana esplanade from near the Charles Bridge towards the Smichov area of town where Ivan had lived as a child in an apartment overlooking the Vltava. The chestnut trees were mainly green now and couples were rowing on the river. There were special spots which they knew where there was a broad view across to the other bank, taking in the National Theatre, Charles Bridge and the esplanade lined with elegant 19th-century houses extending up the river. Ivan liked to think that his parents had once courted at these very spots. For all the brusque drive that he projected, he was capable of misting up at such nostalgic flashes.

The apartment where Ivan had spent his first few years once covered the whole floor when his mother's parents had lived there since the 1920s. But after the Communists took over in 1948 the city government had divided it up and moved in two more families. It was a dreadful blow for his elderly grandparents but there was nothing to be done. There was no escaping punishment for having once been well-to-do.

On these walks Ivan explained his vision to Maika. One topic he described was the possible connection that he saw between the lacunae that Philip had described at Mt. Wilson and a trait that he had observed in a very few colleagues at MIT and then in his research company. They were engineers and scientists who had a peculiar ability to see things differently from everyone else. They were not in general those with the most deductive skill but they had an ineffable talent for seeing what was obvious but was missed by everyone else. Ivan thought that, along with deductive and inductive skills, this separate skill deserved more attention if one objective of science was to identify and investigate lacunae in our knowledge.

Maika liked that, on the one hand, his plans connected to progress in science and might perhaps even contribute to healing divisions between Americans of different outlook. That was laudable and grand. But in her modesty and good sense she appreciated also that these lofty goals were leavened by a connection to their family background of moderation and pragmatism. She remembered reading that the Catholic Church had reported the lowest incidence of miracles in the Czech lands. Privately, she didn't expect any here either.

Chapter Forty-Four

As it happened, plans to push ahead with Ivan's vision had been delayed until the fall in part by issues in his business and also some medical problems. He didn't want to commit to any philanthropic gesture at the Observatory until he had sold the firm and was able to assess his financial position.

The stress may have aggravated a heart condition that had been recognized some years earlier and his cardiologist had recommended taking it easier. The doctors could not say for sure that stress was to blame but Maika insisted that he take the summer off and let his partners run the business. "You can keep in touch perfectly well from Georgeville," she insisted. "There's no need to brave the summer heat here in Cambridge."

So in June, Ivan and Maika packed summer gear into their new Bentley SUV and headed up to the summer place in Georgeville they owned on Lake Memphremagog in Quebec. They drove up the Interstate through Concord, New Hampshire, where their sons had attended school and continued on through Franconia Notch in the White Mountains. The looming crags of Cannon Mountain and Mt. Lafayette were particularly impressive in the late afternoon sun.

They had followed this road through New Hampshire, and its sister Vermont Interstate that ran a bit west, hundreds of times in all kinds of weather. Ivan never tired of the trip. He thought that its curves wending through broad valleys, narrow gorges, past white clapboard villages and high mountains, must have been designed by highway engineers with a rare esthetic sense.

The gentle light green of the budding trees set a hopeful mood in spring. The shimmering heat of summer lent an iridescence to the landscape that recalled to Ivan days of childhood boredom lying in a meadow watching the summer cirrus clouds chase across the sky. The bright red and yellow fall colors were legendary, almost overpoweringly so. Winter brought a hard-edged beauty of sunshine on glistening snow or sometimes, blinding snowstorms. He'd seen other lovely highways in the Alps and elsewhere in Europe, but these two New England Interstates seemed to him amongst the best examples of American art.

A few miles after the border at Derby, Vermont, they left the Quebec highway and followed winding roads towards the sleepy hamlet of Georgeville. Their place was an old white farmhouse with green trim on some twenty acres overlooking the village. The view westward across the lake to Owl's Head Mountain and up towards Gibraltar Point, was spectacular.

Lake Memphremagog had always been a special place to Ivan. Long and narrow, its deep and crystal waters stretched thirty-two miles from the town of Magog, Quebec, at its northern end to Newport, Vermont, at its southern extremity. When he was little his parents occasionally drove down from Montreal for a weekend or a bit longer to Vale Perkins, a little cluster of summer cabins nestled in the shadow of the imposing Owl's Head Mountain on the west side of the lake near the US border. They stayed at the original Perkins homestead, a somewhat run down 18th-century farmhouse still owned by the Perkins family. Old Mrs. Perkins provided a hearty breakfast and dinner.

But the high point was the stories she told after dinner around the fire, of the Indians she knew as a young girl. "Back in my day," she'd say, "the Indians from the settlement at St. Francis would paddle down the lake in their canoes in the spring and sell handicrafts along the way." The details of her stories were a bit of a blur to Ivan since he had been pretty young. But the atmosphere of the old house backed by the glowering cliffs of the steep mountain and fronted by the dark depths of the lake, left a deep impression on a young boy's mind.

Many years later when he was a student at McGill University, he drove down one day to refresh his memory of the place. The bumpy and twisty two-lane road that his father used to navigate in their blue 1950s Ford had been replaced by the Eastern Townships Autoroute, a smooth, four lane affair. Vale Perkins had dwindled to a depressing collection of dilapidated shacks. The Perkins farm was still there but it had been sold after Mrs. Perkins had died and the bartender in the local dive told Philip that her son had left for the city and was driving a cab in Montreal. The experience taught Ivan to think twice before returning to special places of his youth.

When their boys were young, the Vesely's used to come up to Georgeville for a few weeks in the summer. The boys kept busy playing Indians and making bows and arrows and tomahawks. Ivan passed onto them his fascination with Indian lore. He told them how the area around this lake was the Abenaki tribe's winter hunting ground and in the summer the Indians migrated by canoe to the New England seashore to escape the flies and to fish. The irony that their family migrated in the opposite direction was not lost on them.

Maika taught them to look for the boletus "hřiby" and the "lišky," or "fox," mushrooms she used to pick as a child in Czechoslovakia. Their favorite food was boiled beef and dumplings with a wonderful mushroom cream sauce that still smelled of the woods.

Ivan had a little observatory built and spent hours peering at the various astronomical objects that he had observed when he became interested in astronomy as a child in Montreal. Maika and the boys occasionally wandered over but they preferred to play board games after dinner. They accepted that there was no TV and the Internet didn't exist back then.

Winters they skied at nearby Jay Peak, Mt. Orford, or Burke Mountain. When the ice was thick enough, a road was plowed across the lake from Georgeville to Knowlton Landing and it was closest to ski at Owl's Head. When it was too cold to ski downhill they toured the cross country trails or snow-shoed trying to recognize the tracks of wild animals in the snow.

When the lake was calm on hot summer days, as a special treat, Ivan took them out in his pride and joy, a gleaming, varnished Italian Riva inboard runabout that he kept in a marina a few miles up the lake in Magog. The boys screamed with delight when they were little, as he gunned the Maserati engine and the boat shot down the lake, especially when they were allowed to honk its Italian air horns that played

"Funiculi, funicula." Ivan had loved those days; they fulfilled the vision he had held in his mind ever since the first time he saw an exquisite Riva rumbling up Lago di Como in the Italian Alps.

When the boys got older and came home for the summer they found this exuberant flamboyance embarrassing and Ivan had to persuade them to come out in the Riva with friends from school whom they had invited up to the lake. Now, they were both married and had little time to join him and Maika in Quebec. It was hard to tear the grandchildren away from their smartphones. The Riva sat unused and Ivan thought he might sell her.

Chapter Forty-Five

Anna and Igor saw a lot of one another over the next couple of weeks. One night, Igor was able to get tickets to the Ballet, another time to a production of Chekhov's *The Cherry Orchard*. Anna liked him, but it surprised her that he seemed to be able to spend time with her even during the day. She didn't want to press him on it and thought maybe he was on vacation or even between jobs and embarrassed to admit it.

After the Chekhov play Igor said that his apartment was just a few minutes' walk from the theatre and repeated his invitation of a night cap. Anna was feeling mellow from the wine they had consumed during the intermission, and she accepted. She stayed the night and enjoyed herself. Igor was experienced and even over breakfast he made her feel natural and not at all embarrassed. When they parted afterward it was she who asked when they could meet again. Igor told her that he'd like her to meet some of his friends and would get in touch when he had news on that front.

Anna didn't hear from him for a few days but finally he called and said that he and his friends were planning a fishing trip to a dacha belonging to his uncle on one of the lakes about an hour's drive out of the city. He asked whether she'd like to come, it would be just for the day if they were able to leave early in the morning.

On the appointed day Igor picked her up around six AM in a car that he'd apparently borrowed and they sped off. It was grey and rainy but Igor said that was perfect for fishing. She asked where his rods were but he told her that he kept them at the dacha. He told her that she would meet three friends, two men and a woman, whom he'd known for many years. Anna was still groggy from waking up so early and she fell asleep as Igor piloted the car over the narrow and bumpy road.

She woke up just as they were turning into a little lane leading through the forest and ahead she saw a modest shingle structure with trim in the carved wood style seen on many Russian country cottages and churches. There was another car already parked next to the house.

"Wake up, my lovely," said Igor cheerfully, "we have arrived." Anna unfolded herself from the small car and brushed her hair a little to look presentable, checking the effect in the car mirror. She wanted to make a good impression on Igor's friends. The dacha's door opened before they reached it and a portly, middle-aged man greeted them with a broad smile. Igor introduced Anna to this man, Sergei, and asked about the others. "They have taken a brief walk to look for mushrooms," Sergei explained with a conspiratorial smile, glancing at Anna. "Come, let's have a cup of tea to refresh you after your trip." She was surprised that he knew where to find the tea and condiments if the dacha belonged to Igor's uncle.

They talked for a while about the weather and the fishing and as these topics were beginning to wear thin, they heard voices outside and a younger man and woman joined them. After introductions that seemed a bit awkward to Anna, Sergei offered tea to the newcomers and they all seated themselves on a plush sofa and three armchairs arranged facing a fireplace that looked unused. Igor jokingly asked the newcomers, Dana and Evgenii, about their success looking for mushrooms, and everyone sort of laughed.

Then Sergei turned towards Anna and said, in a more serious voice that they had not brought her here to discuss fish or mushrooms. "We are here for a more important matter," he said in an earnest but not threatening way. "You are an important person to us, Anna, and we have been aware of you before you even came to Russia. We know that you graduated with honors from Harvard in astrophysics and biophysics, that you were chosen for a NASA internship, and even of your unfortunate incident with the Congressman. I will not go into detail how we know but the FSB has a large staff of specialists who follow the Washington scene very closely. We are with that organization."

Anna felt faint and stammered out a request for a glass of water, which Igor jumped up and brought her. Sergei reassured her that she had nothing to fear, that they only had some questions for now and they might ask her for a favor at a later date. Anna's mind was racing over the events since she'd met Igor and it all fit. He was a KGB—now called FSB—agent who had been instructed to befriend and seduce her for some purposes that the Russians had in store for her. Once again she felt angry with herself for being so gullible.

After giving Anna a chance to compose herself, Sergei told her that Dana and Evgenii knew her case file best and were most qualified to get her views on certain issues. Dana began with an overview of Anna's record at Harvard, her internship with Philip, and her later time in Washington. She showed clippings of her with Jock at the party and she even had telephone transcripts of her conversations with Cecily about how to handle the rape. Anna was stunned. She asked herself: How did they get this material?

Evgenii continued, informing her that the KGB had operatives working within both the fundamentalist and liberal movements to foster division between Americans. "Creating difficulties for NASA missions like COST is just a small by-product," Evgenii explained. "Our broader interest is to foster divisions between Americans."

"We want you to understand, Anna," interjected Sergei, "that all this work is to promote peace and harmony in the world. America is simply too rich and powerful and Mother Russia needs help from honest and dedicated people like yourself to help balance the scales. That's all we ask."

Anna tried to steady her voice to ask what favor they expected from her. Sergei answered in a measured but kindly voice that they would discuss that in their next meeting. "For now, I want you to just let this settle in your mind. If you are willing to help us in our quest for peace you can be sure of a handsome reward, both to yourself and to your grandparents. For the time being you must not try to leave the country or even communicate with anyone about our meeting or anything that was discussed. Especially not with the US Embassy."

With this, they all stood and prepared to leave. The others left first. Anna asked to use the facilities and looking into the bathroom mirror was shocked to see that she looked like she had aged ten years. Even her vivid green eyes had lost some of their luster.

Chapter Forty-Six

Nowadays Ivan and Maika mostly went up to Georgeville on their own. Their sons no longer lived in New England and few of their Boston area friends cared to make the four-and-a-half hour trip. After years of swimming in the lake at the village landing they decided to build a horizon pool. Ivan also oversaw the construction of a clay tennis court that would be easier on their knees than a hard surface.

When Emily called in late July asking about Ivan's health, Maika invited them up for a few days. She suggested that they might want to make it part of a trip to Montreal, which is only an hour and a half farther up the road. Ivan spoke with Philip later and told him that they could use the opportunity to discuss what Ivan had in mind with the CFA.

They arrived on an afternoon mid-week to avoid the traffic and went for a swim to cool off and recover from the drive. The black fly season was over but mosquitoes hovered expectantly. Maika explained that this was the season when the abundant dragonflies were devouring the remains of the mosquito population, but they had some ways to go.

She told them that these lovely insects reminded her of the jet helicopter that Ivan used to fly them up in when they first acquired the property. "He enjoyed learning how to fly it and the boys loved the ride, but the neighbors here complained of the noise and the border crossing became more difficult when security was tightened after 9/11." She suggested then that while they were waiting for the dragonflies to finish their job, they might move into the screen porch to enjoy some iced tea.

On the porch Philip brought Ivan up to speed on the COST project and his continued participation in the scientific planning while Emily and Maika discussed their children and Maika and Ivan's fourth grandchild. They agreed that this younger generation seemed to be making child rearing harder than it really needed to be. "I would never have dreamed of asking each of our boys what he wanted to eat," said Maika. "I just cooked a meal and everyone ate it and said: Thanks, Mom," Emily agreed. "Our kids were making their own peanut butter sandwiches for lunch since they were in elementary school."

Ivan's tennis had become more cautious because of his heart condition but they still enjoyed a good two sets before he apologized for having to quit. After another swim to cool off they adjourned to the porch for some gin and tonic. "We have made dinner reservations in a nice bistro we have here in Georgeville," said Maika.

Over drinks, Ivan told them that bistros like this one were a benefit of the ascendancy of the French in Quebec since the late 1960s. "When my family came to Montreal just after WWII, the gastronomic scene was about as bleak as it was in Boston at that time. In Quebec the common denominator was hot dogs and hamburgers, just like in the States. But beginning in the mid-'60s young Montrealers discovered French cuisine and restaurants serving very good food sprang up everywhere. These were soon joined by a wide variety of other ethnic foods."

Their experience at dinner confirmed this and when they drove back up to the farmhouse Ivan said that the weather was expected to be calm tomorrow, and he promised to take them on a trip down the lake in the morning before any wind came up. It was a bit early to turn in yet, so they settled in on the porch for a little brandy and Ivan told them something of the local history.

"This area was settled first by Tories fleeing New England during the American revolution," he began. "Georgeville's original name was Kopp's Ferry after a Massachusetts family that settled here in the late 1700s and ran a ferry service across to Knowlton Landing. For many years that was part of the stagecoach route between Boston and Montreal. There was little French settlement here because the original property grants bestowed by the French King in the 17th century were narrow strips of land fronting on the St. Lawrence River and extending only a few miles inland. In this hinterland farther from the St. Lawrence it was Indian country. The

British who captured Quebec from the French in 1760 at first welcomed these United Empire Loyalists as they were called here.

"But later they began to worry that these transplanted Americans might not remain loyal to the Crown so they encouraged immigration from the United Kingdom and also of French Quebecers. These French had shown a preference for supporting the Crown rather than the Americans who tried to attack Quebec during the American Revolution and also during the War of 1812. Canada and the US were not good friends back then!"

After the Frobishers retired that night, Emily lay awake listening to the sounds of the night. She recognized the strange, distant shrieks of distress that pierced the silence from time to time as the cries of the beautiful black and white mottled birds called loons on the lake below. She had heard them a long time ago as a young girl at Camp Mudjekeewis on Kezar Lake in Maine.

She thought about how the loons had been here through all those millennia before Europeans arrived, when the Indians had these woods to themselves for the ten thousand years and more since glaciers retreated and left behind Owl's Head and this magical lake.

Chapter Forty-Seven

Emily awoke early and had a swim before breakfast. The mist was still on the lake below and the mountains were resplendent in the morning sun. She dried off and sat on the porch admiring the view and waiting for the others to wake up.

After breakfast Maika said that she would stay to prepare some lunch and dinner, so Ivan and the Frobishers set off for Magog on their own. The scenic road ran near the shore and Emily marveled at several impressive estates that they passed on the way. "They belong, or in most cases once did, to wealthy Montreal families like the Anguses and Van Horne's who built the Canadian transcontinental railways in the late 1800s. The Molsons' place is down the lake nearer to us. They used to own large steam yachts and entertain British aristocracy here. Montreal Anglo society has always been more British than the British," he told them with a smile.

Ivan's Riva *Amica* was ready for them when they arrived at the marina, shining in new varnish and polished chrome. "Maintaining this boat costs a fortune," Ivan confessed, "but as the owner of the boatyard cheerfully reminds me, you can't take it with you! She's well known up and down this lake and I've had offers, but so far I've put off selling. Thanks for providing the motivation for my first ride this summer!"

The Frobishers boarded gingerly, afraid of disturbing the beautiful mahogany or the exquisitely soft red leather seats. Ivan laughed and encouraged them by jumping in himself and starting the engine. The low rumble of the Maserati engine was already exciting for his guests. "The Italians are geniuses at muffler design," Ivan commented, seeing their admiration.

Ivan put *Amica* in gear and they carefully made their way out of the marina. He waved to some admiring onlookers and let the airhorns sing out as they passed the last spit of land. As he advanced the throttle the engine sound rose from a low rumble to a roar and the boat leapt forward. "Hold onto your hats," Ivan yelled, "here we go!" Emily and Philip felt a thrill as the boat accelerated towards 50 mph, shooting along the shorelines of the large estates they had glimpsed from the road. Occasionally Ivan throttled down as they approached wakes of other craft.

"The original Rivas had flat bottoms that sort of worked on the Italian lakes," Ivan shouted, "but the American yacht designer Dick Bertram persuaded the old man that changing the hull to a deep-V cross sectional shape would improve the boats' performance on the Mediterranean where there was a potentially larger market; this boat is one of the newer design."

Philip held on tight as they rocketed down the lake past the headland of Gibraltar Point and along the Georgeville shore to beyond Owl's Head before Ivan put them into a sharp banked turn as they approached the US border and headed back. "The stricter rules enforced by Homeland Security make crossing over into the US more difficult," he shouted over the engine roar. "We could go to the end of the lake at Newport, Vermont, but the Canadian part is more scenic anyway."

Emily enjoyed the speed more than Philip and looked rather dashing with her print scarf flying when he glanced her way to see how she was faring. He didn't relax until they slowed for the approach to the marina and tied back up at their slip. "I hope you liked that," Ivan offered, smiling and looking for their reaction. "Let's head back and have a swim."

Maika had prepared a light lunch and joined them in the pool beforehand. It wasn't too hot or buggy so they ate outside on the patio. She proposed that they all take a rest afterwards and reconvene for a walk around the property in an hour or so. "I will see how successful you are at hunting for mushrooms," she said, jokingly. "I need some fresh ones for my breakfast omelets tomorrow morning."

After a nap they applied some insect spray and filed across the meadow where the house was set, following an overgrown path into the woods. All about, daisies and black-eyed Susans nodded in the breeze.

Philip noticed that the relatively short firs and spruce had the appearance of an almost boreal forest. Here and there they saw stands of birch, poplar and a few maples. But he remembered that the limit of northern extent of oaks lay about a hundred miles to the south.

"Mushrooms of the boletus variety often grow on the edge of the woods, so keep your eyes peeled," Maika announced. She carried a little wicker basket to bring the prizes home and showed them a picture of a boletus that appeared on the lid. "Some days I fill the basket in half an hour, on others I come home empty-handed," she explained. "You never know."

On that day their haul was only one, but it was Emily who spotted it first in the moss under a little fir tree. It was a beauty, a textbook "hřib." Its head glistened in shades of burnished mahogany and the nice white stalk showed a little black mottling. When Maika carefully plucked it out of the moss and sliced the end off with a sharp pocket knife, the stalk had only a few worms at the very bottom and otherwise was quite pristine. Emily was even more excited by her find than by the boat ride. "It's a good size and will be plenty for the omelets I'll make you for breakfast before you head for Montreal tomorrow morning," Maika assured her.

They were ready for a swim after they returned warm from their walk. Then it was time for some cocktails on the porch since the weather had turned more humid and the mosquitoes were eyeing a potential feast by the pool. "We'll have a thunderstorm tonight, I'll bet," Ivan predicted as they hurried back to the house to take cover from the bugs.

Chapter Forty-Eight

Dark clouds were gathering over Owl's Head as Ivan mixed their drinks. "I've ordered gin and tonic in places all over the world and been amazed at how rare it is to find a decent one outside the US," he told them. "Even in England and India, where they were invented, more often than not you get two little ice cubes and some limp old tonic that was opened a month ago," he quipped. The Frobishers nodded and added their own sad experiences as they savored Ivan's frisky version, adorned with a quarter of a lime on top.

For dinner Ivan and Maika had planned a campfire to roast real Czech "buřty" sausages like knock-or bratwurst over the fire. Ivan had made his house special potato salad peeling and cubing the potatoes boiled not too soft, adding chopped Polish dill pickles, diced red onion, a can of green peas, and mixing together with plenty of mayonnaise, some white vinegar, salt, and pepper. "It's important to put all the ingredients in a good-sized bowl after the potatoes have cooled and only mix once," Ivan explained. "If you mix too often after adding each ingredient, the carefully cubed potatoes will turn to mush."

While Ivan and Philip prepared the campfire in a ring of stones that looked like it had been well used over the years, Emily helped Maika carry out some ears of local corn for grilling, the various hot and sweet mustards, and other elements of the feast. Ivan joined them carrying a few cold bottles of the Pilsner Urquell beer that they had enjoyed in Prague. The mosquitoes seemed to be held at bay by the smoke wafting from the fire.

Threatening dark clouds lent drama to the sunset as Ivan passed out sharpened sticks and showed them how to notch the sausages in a cross pattern at each end. "That enables the sausage to open up as you roast it, so the inside gets as well cooked as the exterior." He then held his stick over the fire, rotating the sausage regularly. The Frobishers followed suit and soon were cheerfully tasting the fruit of their labors, braving the swirling smoke and searing heat they encountered if they moved their seats too close to the flames. Maika had placed the corn on a grill that she briefly held over the flames. "My mother used to wrap potatoes in tinfoil and place them in the embers but I never really liked the charcoal encrusted product that emerged," she confided.

As they were preparing to roast a second round of sausages the breeze rose and the first drops of rain sizzled in the fire. At the same time a jagged bolt of lightning pierced the darkness on the other side of the lake. "Maybe we should move indoors to the porch," Ivan recommended. So they hurriedly packed up the condiments and utensils and headed indoors. Ivan put on a rain jacket and stayed out a while longer to finish roasting a few of the sausages that were not yet done. Then the downpour began in earnest and he also beat a retreat.

They finished their meal in comfort on the screen porch watching the drama of lightning and thunder and violent gusts of wind outside. For dessert Maika surprised them with slices of delicious Czech "ovocny koláče," an open-faced fruit tart made of yeast dough with seasonal apricots, peaches or berries topped with a light crumble of sugar, flour and butter. "Memorable," exclaimed Philip. Afterwards, when Emily followed Maika into the kitchen to help clean up, Ivan suggested to Philip that they stay awhile to discuss the project that Ivan thought he might be able to help with.

"You mentioned the surprising gaps or lacunae that exist in science," he began. "Questions that most of us don't see although we realize after they are pointed out to us that they have been staring us in the face. After our conversation with Abe at Mt. Wilson Observatory I've been thinking that the question whether our brains are capable of grasping the universe is a glaring example of such a lacuna. Who is making a serious attempt to study that?

"Even more importantly, it seems to me that the role of faith in both religion and cosmology may offer a pathway to reconciling the fundamentalist Christian right with the liberal left. That is a topic that Abe touched on at Mt. Wilson and I see great potential there. What is needed is leadership. Progress will require individuals of unimpeachable liberal credentials to point to the conundrum facing cosmology, on one hand. It will also require fundamentalist leaders like Southern Baptist ministers to join the fight against global warming. The most powerful tool to reconciliation is to identify areas where both sides can show that they stand by their principles even when it hurts.

"What am I willing to do to help? I am proposing to provide seed money for an Institute that would focus on such activities. I'm prepared to give $15M over three years to Harvard, if you are willing to act as Director. I envision the Institute including cosmologists, psychologists, philosophers, neurologists, and theologians. I'd require that the University and the Federal agencies like NASA match my donation."

Philip was taken aback by the focus and scope of the proposed project. "Wow! That's a lot to think about," he exclaimed. On a practical level he immediately recognized the difficulty of getting Harvard and the agencies to commit to this project. Ivan saw his hesitation and divined the reason for it. "Let me add that my proposal could be an important tool for getting COST funded," he added, looking at Philip intently.

"In terms of dollars and cents investing a few million in helping to grease the ways for a $20B mission is a wise strategy for NASA. For Harvard, the prospect of a return in the form of at least hundreds of millions in hardware contracts to the Center for Astrophysics should also warm the heart of the most dour of comptrollers."

Philip nodded. He now saw the broader attractiveness in this modest proposal. Cecily would be intrigued, he knew.

Chapter Forty-Nine

As soon as they emerged from the dacha's drive onto the main road Igor turned to Anna and said, "I know that you have every reason to distrust me but please listen to what I have to say. The favor that the KGB expect from you is that you will renounce your statement that your experience with the Congressman was consensual. They want you to write an exposé in which you tell the whole story of how you were pressured into that statement to save the reputation of NASA and Harvard, and how in fact, you were planted in the Congressman's office by Cecily Thomas and Philip Frobisher to discredit Jock White and the fundamentalist lobby.

"They will either release this story to the world media, not only harming the reputation of NASA but suggesting dirty dealing and corruption more generally at the highest levels of US science, or they could just store it along with the many other bargaining chips they keep in their files for a time when they need a favor from the FBI.

"In either case, you need to get out of Russia immediately. If you don't you will be left with the option of either acceding to their demands and becoming a pariah in US scientific and political circles or refusing them and taking the consequences."

Anna asked what consequences were likely. "I doubt that they would do anything too dramatic because that could trigger an international incident that might go against them. But they could hold you here for as long as they wished, in less than comfortable conditions."

"Would there be any repercussions to my grandparents if I get out illegally?" she asked. "I doubt that because, you may not be aware, your grandfather is a Hero of the Great Patriotic War. He saved the life of an important Soviet general during the siege of our city by the Germans by retrieving and throwing back a live grenade that had landed in their jeep. That is why your grandparents have a relatively nice flat and enough of a pension to feed a guest like you.

"I am the person entrusted with watching your movements so we have some latitude in how we go about getting you out. I do not recommend trying a legal crossing because they have certainly issued an alert to stop you at the border. I have given this some thought over the past weeks and I have a contact on an excursion ship that leaves here tonight. He can hide us in the engine room overnight and the boat will dock in Helsinki early tomorrow morning. I recommend that we drive directly to the harbor and get onboard."

Anna was puzzled by all this. "Why are you doing this for me and what do you mean by 'us'? Are you planning to come along?" They were passing through a forested area with no traffic and Igor eased the car onto the shoulder of the road.

"Anna, this is a very bad time to be telling you this, I know, but I love you. I would have pursued you even if all this had not happened. I understand perfectly that you are wary that this is just a trap, and I wish that I could prove my sincerity, but please, for your own good, trust me.

"I am accompanying you because I love you and would like to have a life with you. Even if I didn't feel that way, I'd have no choice. If the KGB suspect that I helped you escape, that would be treason which is punishable by execution."

Anna was silent. She felt powerless and could not see any way of testing whether she should trust this man who was able to smoothly befriend and seduce her without so much as missing a beat. They drove along in silence through the afternoon with the poplar-lined country road elevated above the surrounding fields gradually giving way to the outskirts of the city. Staring out the window, she was dimly aware of people going about their daily routines and wondered how they could be so oblivious to her predicament and also, she now began to comprehend, to Igor's.

As they approached the city center Igor reminded her gently that she needed to decide. "We are approaching the intersection with the road to the harbor," he told her calmly. "Should I turn off there?" Anna's stomach churned, and she felt light headed and thought she would pass out.

"Please stop the car and let me get out to clear my head," she asked him in a faint voice. Igor saw a McDonald's ahead and turned into the parking lot. "Let's go in and have a Coke," he said, trying to relieve the tension. Anna smiled weakly at the irony of making a life-or-death decision over a Coke at a McDonald's in St. Petersburg, Russia.

They sat in a booth like two teenagers dating and Anna remembered that she hadn't eaten all day. "I'd like a cheeseburger with fries, if we can spare the time," she pronounced more emphatically than she felt. Igor got up and ordered it for her at the counter. When he returned and they were waiting for it to be delivered, Igor put out his hand to her and after some hesitation she took it. He cupped her hand in both of his and kissed it. "You must believe me, Anna, I have been in love with your beauty and grace and kindness to your grandparents since I met you. I have never stopped thinking about you. I cannot imagine life without you."

Anna fought to focus but her mind was filled with random thoughts. A wit at Harvard had commented to her once that "if you can fake sincerity, you've got it made." She looked at Igor and wondered whether that applied to him. But she detected that he wasn't as self-assured as he had been before this trip to the dacha. When the burger came, he asked her sheepishly whether he might have a bite, that he should have ordered one too. He also had trouble opening the packaged condiments and squirted mustard all over the table. She didn't have much to go on but it seemed that he was as nervous as she was.

"Fine," she said finally after they had eaten the last fry. "I'm feeling better now. I'll agree to your plan. Luckily, I have my passport since I'm in the habit of carrying it everywhere when I'm in Russia." Igor said that he had his papers too and some money. "I keep part of my savings hidden in my apartment; no one here trusts Russian banks." As they exited the McDonald's Igor took Anna by the shoulders and turned her towards him. They kissed and Anna believed more than before that his feelings were real.

Igor turned out of the parking lot and drove towards the harbor where they saw the giant cruise ship looming over some smaller craft. It was quiet on this part of the docks because the passengers were out doing the town and wouldn't be boarding until much later.

Igor drove the car alongside the ship and turned so its headlights could be seen from onboard. He blipped them and they waited. In a few minutes a small door opened in the ship's side and a gangplank was extended. A man beckoned to them and they crossed. He and Igor exchanged some muttered words and Igor passed him an envelope that the man opened and counted the bills. He led them down several metal staircases into the depths of the vessel and opened a door into a small room lined with lockers with a metal table bolted to the floor and hard benches along two of the walls.

Their host informed them that he would lock them in this room; no one had any business disturbing them until next morning. "You can sleep on the benches if you like and use the facility here through this other door. The water there is drinkable and there are some paper cups also. I'll come and get you when we arrive in Helsinki. Good luck!"

Chapter Fifty

Philip outlined Ivan's proposal to Cecily after he returned from Montreal, cautioning her that it was contingent on the successful sale of Ivan's business. This dragged on through the summer and fall, but in November Ivan told Philip that it was imminent and he was hopeful that he could follow through on his offer.

During this period of waiting Philip had mulled over the offer and had come to like it more. He began to see it as perhaps the opportunity he had been awaiting to discover something truly important and meaningful to do with the rest of his career. Bob Leighton's admonition—that as you get more senior you can pull off things that a younger person could not—came back to him.

Cecily was clearly intrigued but she thought, as did Philip, that Harvard's willingness to provide a home and a considerable amount of its own funding, would be key to persuading her boss to contribute matching NASA funds. She asked Philip to draft a proposal with Ivan that she could take upstairs.

When Ivan told Philip after Thanksgiving that the sale had gone through, they agreed that it was time to sit down with Nancy Stein and Cecily and perhaps the Harvard Dean of Science, to air the proposal and address its feasibility. Cecily needed to carry out a site visit for a NASA project underway at MIT so she proposed to Philip that they meet in Cambridge in December.

Brian was doing his best trying to work the influence that the aerospace industry had on Congressmen in the states most likely to benefit from the COST mission. But they were distracted by the even bigger pot of money connected with the plans of the NASA Manned Operations

Division to send astronauts to the moon again and even to Mars. Without support from the Christian lobby COST seemed to have little chance. Protests from the science community that such manned missions had little scientific value fell on deaf ears.

Her personal relationship with Brian had cooled when she found out that he hadn't been entirely candid and his divorce was still a work in process. Seeing him socially was already chancy enough without the danger of being named in court as a participant in his adultery. That could attract undesirable attention to their tricky NASA–contractor relationship.

It was snowing lightly when Cecily arrived in Boston. As her cab drove her from Logan Airport along Memorial Drive to the hotel on Harvard Square, she looked out at the Charles river, dark between its banks glistening with snow. During summer and fall visits to Cambridge she had enjoyed the sight of sailing dinghies from the community boating program flitting about in the sunlight on the lower basin. Upstream the river was taken over by numerous rowing shells. Now she saw only MIT and Harvard students hurrying along the riverside sidewalk, coat collars pulled up to shield from the biting wind.

After dinner at the hotel Cecily went out for a walk around the Square. All about her Christmas shoppers scurried between stores, shuffling cheerfully through the first couple of inches of snow. The wind had abated and the last flurries settled gently on the ground in the little park she crossed. The air was bracing but not too cold and she enjoyed the lift it gave to her spirits. The Christmas season cheered her but also brought some sadness. Once again, she'd be celebrating the holiday with just her mother and at the same annual office parties.

In the months since Anna's departure for Russia, Cecily had wrestled with the unpleasant fact that, whatever her plan for getting around the fundamentalist opposition to COST, she'd have to deal with Jock White, a man who had raped at least one and perhaps more interns without suffering any consequences. She found that prospect repulsive.

So she was astonished to learn, just before leaving for Cambridge, that Jock White had been killed by a cougar while hunting in Idaho. His guide told the media that he had never seen anything like it in all his many years of guiding. His death was widely covered by the media with emphasis on the good he had done for his constituents. Then the attack became the subject of controversy between Idaho ranchers and the animal rights lobby. She felt sorry for his family but was relieved that she wouldn't have to deal with the man.

Walking through the snow and looking into brightly lit shop windows helped Cecily clear her head. She stopped into a little chocolatier for a decadently delicious hot chocolate. It warmed her and she decided that she would tell Philip about Anna tomorrow. She felt confident that she could deal with any consequences that this revelation might produce.

Chapter Fifty-One

Next morning after breakfast at the hotel, Cecily walked up from the Square to the Center for Astrophysics since Ivan had arranged with Philip that they would meet in his office. The winter morning sunlight illuminated Philip's corner office in the Wolbach Building with a rosy glow. Cecily was the first to arrive and despite her good intentions last night, she was relieved that Philip was preparing for his observing run in Hawaii and had other things than Anna on his mind.

As the others filtered in, Philip introduced Cecily to Nancy Stein, Ivan, and the Science Dean. Philip was wearing his usual winter garb of corduroy trousers and wool turtle neck, Cecily had appeared in her NASA executive uniform of well fitted grey pants suit, while Nancy, who had no sense of clothing, wore a clashing ensemble of skirt and vest. The goateed Dean came dressed in an all-black outfit of trousers and cotton mock turtle with an exotic peace pendant around his neck. Surprisingly, given the weather, he was shod in yellow sneakers.

Against the background of this motley haberdashery, Ivan stood out like a peacock amongst a covey of pigeons in a beautifully tailored Italian blue suit and carefully polished, two-color shoes. Even before anyone had said anything, he had his academic hosts perplexed and back on their heels.

To begin, Philip asked Ivan to go over his proposal, referring to the document that he had written and distributed to the others present a week earlier. He elaborated essentially on the points he had outlined to Philip in Georgeville and for the benefit of Nancy and the Dean he stressed the importance of Harvard's involvement to supply credibility while alluding to the potential financial benefits.

For Cecily he stressed the ways that the Institute could help to mitigate fundamentalist opposition to COST by opening a dialogue with that community about the roles of faith and scientific method in understanding the universe and modern science more generally.

There was little comment during his presentation. Afterwards they all thanked him for a clear exposition and wanted to know whether any of the potential members of the Institute's staff had been consulted. "So much depends on the quality of the faculty," commented the Dean.

"It seems to me that the only sure way to assess that is to solicit the reaction to this proposal from a sample of those whom you would wish to attract," he continued. "Of course, the best approach is to pull in one or two heavies in the key fields. If past experience is any guide, the others will follow."

Cecily agreed and added that high salaries would be essential for those key first hires. "Academics play coy about money but you'd be amazed how often big guns are pulled out of endowed chairs at the Ivies by dangling big salaries before their eyes . . . or by offering their spouse a faculty position."

Nancy changed the subject asking whether Ivan had in mind an Institute of bricks and mortar or rather a virtual entity in which staff might be spread around the world but be joined by the Internet. "I consider it essential that the core staff be located in the same building. That sounds ambitious, I know," he answered. "But when Harvard's new science campus in Allston is complete in a few years there should be space either there or in the older buildings closer to the Yard, which should be freed up."

Philip suggested that, with noon approaching, it might be time to stop there and give everyone a chance to process what had been said and get comments from colleagues. "I encourage discussion of the general concept with the community," he said, "but it would be premature to approach potential key staff. I want those approaches to be made by a panel including Philip and me. We'd need to agree on a list and a procedure to follow."

They all agreed and thanked Ivan for proposing such an interesting project and offering to provide the lead funding. Philip concluded saying that he'd act as clearing house for any feedback and would suggest a time for their next meeting.

Chapter Fifty-Two

After the Dean, Nancy, and Cecily had left, Ivan and Philip went down to the lunch area in the atrium of the Center for Astrophysics to get a cup of soup and sandwiches to bring back to Philip's office. "How do you think that went?" asked Ivan. "I think that they are being polite because your proposed donation is nothing to sneeze at, but I'm sure they see many hurdles ahead," Philip answered. "Your point that space shouldn't be a problem when the new Allston Science Campus comes on line was well taken. I hadn't thought of that."

"We may have have erred in not inviting the Dean of the Divinity School," Philip added. "To him the donation you are offering would probably be more exciting than for the others." "True," answered Ivan, "but there is something to be said for focusing on the people who are likely to offer most resistance first. If we can persuade them Divinity will come along, I'm sure."

"There are so many donors lining up to give hundreds of millions to Harvard," Philip added, "that the amount in question here isn't in itself enough to get them fired up. Back in the 1960s, Harvard turned down a donor's offer to build a whole visual arts center on the grounds that visual arts were not part of a gentleman's education! But when Yale let it be known that they would be delighted to receive the center, Harvard suddenly thought again and the Carpenter Center has become part of the fabric of life here since that time."

"So are you saying that we should open the proposal to other universities?" asked Ivan. "It might be useful later," answered Philip, "but for now let's keep it simple. I'm sure that Nancy and the Dean will read over our proposal and even show it to some others but they aren't going to provide the initiative to move this forward. That will need to come from us. Cecily also may show it to her boss but then they'll sit on it, waiting for us to make the next move."

When Philip got home that night Emily told him that he'd had a call from a youthful sounding woman named Anna. "She said that you'd know who she was," said Emily, pretending she wasn't interested. "Here's her number, she asked you to call her back as soon as convenient." Philip wondered what that might all be about but said that it could wait until after dinner. When Emily asked who she was, he answered nonchalantly that she had been an intern with him for a while at Harvard and that he had no idea why she was calling.

Emily left for a church board meeting after dinner and Philip rang the number after she had gone. He immediately recognized Anna's voice and it brought back memories that he preferred to have buried. "Anna, how are you?" he blurted out in a rush. "I've had an interesting few months, Philip," she responded quietly. "I'd like to meet and tell you about it." He noticed that she called him Philip now, not Dr. or Professor Frobisher like last year.

"Where are you now?" he asked. "I'm in New York staying with my parents but I'm planning to come up to Cambridge in a few days. Could we arrange a time to meet somewhere we can talk privately without being disturbed?" "Of course," he answered, thinking rapidly where and when he could arrange that. "How about the bar at the Sheraton Ambassador Hotel across Concord Avenue from the Longy Music School? It's usually deserted around lunch time."

Anna didn't sound thrilled at the choice but she agreed that there were few places around the Square where one could hope to have uninterrupted peace and quiet for a serious conversation. He told her of his commitment in Hawaii in a few days and suggested that she come up tomorrow or the day after. "Would that suit you?" he asked. "That's fine," she answered, "the sooner I get away from my parents the better. I'll see you there tomorrow at noon."

After he got off the telephone, Philip struggled with conflicting thoughts. She sounded more self-assured than when he had last seen her a year ago but also seemed weary. He wondered what had happened to her after she left for Washington to take up the internship and felt badly that he'd never asked Cecily. The truth was that he hadn't wanted to raise any suspicions that he and Anna had been more than professionally connected. Contacting her directly would have been out of the question in the crazy atmosphere of gender conflicts that poisoned US academia these days.

When Emily returned she asked whether he'd reached Anna and what the reason was for her call. Philip decided that it would be best to tell her the truth that Anna had asked to meet with him tomorrow for lunch. "Why does she need to meet with you all of a sudden?" She demanded in a concerned voice.

Philip said he didn't know but she was a star student and research assistant and he felt that he owed her the meeting that she was requesting. "Do you want me to come along?" asked Emily. "You never know what nonsense these young girls carry around in their heads. Before you know it their lawyer is charging you with some heinous deed." Philip was able to calm her fears somewhat, but she spent longer than usual watching TV and didn't come to bed until after he'd fallen asleep.

Chapter Fifty-Three

Philip arrived first at their meeting on the next day. He noted with satisfaction that the place was empty and so dark you could stumble over the plush easy chairs arranged around circular black polished tables. He told the waiter that he was expecting someone to join him and asked for a luncheon menu for now. He also asked for a light so he could read the menu and the waiter brought a candle in a glass vase.

Anna arrived a few minutes late and as he watched her willowy shape wend its way between the tables, he found himself as nervous as he'd been on his first date in high school. He had addressed ballrooms full of scientists and dignitaries as a famous cosmologist but now his throat was dry and he had to sip some water before getting up to greet her. He was unsure whether to kiss her on the cheek but she solved that problem by simply putting out her hand to shake.

Philip offered her a drink and she asked for a glass of white wine. He had the same, to keep her company. After they had ordered lunch Anna thanked him for meeting with her. The fleeting smile she gave him and her green eyes still captivated him, he could not deny it. "I have had a very troubling few months, Philip; you deserve to know what happened to me."

She told him the whole story starting with how things seemed to be going well with her internship in Washington and then the incident with the Congressman and how it was hushed up to avoid a scandal involving NASA. She went on to describe her exile in Russia and the incident with the KGB. She told him how Igor had helped her escape and how they had made their way to the US despite the FBI's delays in processing him as a defector from the KGB.

"Igor is still in Washington answering questions and waiting for the FBI to create a new identity for him," she said with a sigh. "I am very grateful to him for the risks he took for me and we love one another. But all this is very stressful and it is hard to be happy together."

Philip listened, shifting his weight and clearing his throat at pauses in her narrative, trying to process what he was hearing. When she had finished he ordered two more glasses of wine. "What an amazing and shocking story! I'm terribly sorry that all this happened to you," he said, looking at her earnestly. "Please believe that I had nothing to do with placing you in the internship with Jock White. I think you'll agree that you came to me asking for a recommendation and the letter of support that I wrote you said nothing about what placement you would be suited for. Do you agree?" Anna nodded in assent.

"I can't comment on Cecily's role in this. It's possible that she did see an opportunity here but I can't believe that she truly foresaw the consequences. I have known her for years and I respect her judgement and her sense of ethics. Maybe she wanted to believe that, if Jock came on to you, you were old enough to handle such a situation yourself." Again Anna nodded but Philip saw that she was less confident in his view of the matter.

"Have you spoken with her since your return and told her what happened in Russia?" he asked. "No, I haven't," she answered. "I'm not sure that I want to confront her with it. The FBI may have informed NASA of the incident by now anyhow." Philip thought to himself that someone should tell Cecily, but maybe Anna was right to stay out of it.

Anna told Philip that it was helpful to her to get this all off her chest. "I can't talk to my parents about any of this, they would freak out if they heard that I had been raped by a Congressman and then almost taken captive by the KGB. Stuff like that only happens in bad movies."

Philip asked her what she planned to do next. "I want to apply to law schools as I had meant to do this spring, but it depends on where Igor gets placed in his new identity. We want to get married but I'm not sure that I will be allowed to communicate with you, for example, if we are given new identities together. For now I'm just visiting some friends but

I'm keeping most of this stuff to myself. I've been warned that the KGB has a wide reach even in the US."

At this, Anna stood up and thanked him for lunch and the chance to talk. "You have always meant a lot to me," she said in a husky voice and placing her arms gently around his neck, kissed him. Then she gathered her coat and left. Philip sat down and pretended to examine the bill, but his mind was elsewhere.

He walked back up Mt. Auburn Street to his office thinking about Anna's story. Cecily's role in this troubled him. Had she knowingly placed Anna with Jock White to entrap him? Maybe the truth would never be known. It wouldn't be the first time, though, that he'd seen senior women fail to protect their younger colleagues. Rarely acknowledged impulses sometimes came into play.

Blessedly he had lots of busy work to do preparing for his telescope run in Hawaii. He was hoping that the weather would be better this time and that they would be able to fill in a few more points on the plot he was preparing to present at a meeting in a few months. Philip was curious to find out whether the missing points trended upwards or down because that would influence interpretation of the universe's mysteriously accelerating expansion.

He hesitated to contact Cecily but felt that she ought to know. Finally, he tried her but she wasn't available so he left a message saying that Anna had visited him and told him what had happened with Jock White. He was surprised, he said, that Cecily had not mentioned that to him. He asked her to call back because she needed to know what had happened to Anna in Russia. He added that he was leaving for an observing run in Hawaii on the day after tomorrow, but she could get him on his cell phone while he was travelling.

On the way home that evening, he thought of what he could tell Emily without alarming her. He had to have some explanation of why Anna had sought him out and decided that he'd tell her about Anna's internship in Washington, that she had been upset by a Congressman's advances to her, and she needed to get the experience off her chest. That wouldn't be lying, just keeping it within bounds that wouldn't keep Emily up all night.

His explanation satisfied Emily and they talked more about what she planned to do during his absence in Hawaii. She mentioned that Maika had asked her to a girls' night out dinner in town and a Gilbert & Sullivan production since Ivan would be away the same week. She wasn't a big fan of driving into the city alone at night, especially now with the roads slippery again but she thought that it would give them a chance to get better acquainted and that she would go.

Cecily called back the next morning apologizing for being hard to reach at meetings all day before. "I'm terribly sorry that I didn't tell you about Anna," she said. "I was going to discuss it with you when I came up to Cambridge for our meeting just now but the opportunity for a private talk never seemed to come up. What happened in Russia and is she back?"

Philip told her the story that Anna had related to him yesterday and for the longest time Cecily didn't say a word, he just heard her breathing. "My God, I had no idea!" she finally whispered. "Philip, I am just devastated by this news. I feel that this is all my fault and I should resign."

"I don't think you should do anything hasty," Philip recommended. "She's safe and she's back. This new identity thing worries me, she has too much potential to be wasting herself hiding somewhere in an Iowa cornfield. But maybe they won't get married, who knows?

"Tell me, Cecily," he continued, "we've been through lots of stuff together over the years, what were you thinking when you placed a beautiful girl like her with that lothario Jock White? Were you trying to entrap him?"

Cecily was silent for a while. "Honestly, I don't know, Philip. When I think about it, I can't believe I'd do anything like that, but it just spiraled out of control. Maybe I thought that it wouldn't come to that, that we could catch him in some lesser misdemeanor like trying to kiss her. She's twenty-two years old and a Harvard graduate after all, not some star-struck bimbo teenager."

Philip thought to himself that he had yet to meet anyone, Harvard grad or not, who wasn't attracted to money, looks, and especially power, but he kept that thought to himself. "As a last thought, Cecily, do you know whether your boss is aware of this whole story? He may have been alerted to the KGB aspect of it by the FBI."

"He certainly knows about the Jock White affair," she answered, "he was in on the decision to kill the story. Do you think that I should sound out whether he's heard the rest?" "He'll find out eventually if he doesn't know already," Philip responded. "I'd recommend telling him." They left it on that note and Philip went back to preparing for his trip.

Chapter Fifty-Four

When Philip returned from his observing run in Hawaii, winter had taken hold in Boston and it was looking like Christmas might be white for a change. The kids were back from school and by and large their news of fall term brought cheer to the season. Becky was doing well academically and brought home a friend for the holidays, a Danish girl named Octavia who had interesting observations about European and American life.

Philip noted that, as had Ivan and Maika, she immediately remarked on the interesting portrait of Emily's grandmother hanging in their entrance way, when she first arrived. None of their American friends had ever paid any attention to it.

To Philip's surprise, William had become involved in the drama club at school and had a minor role in a forthcoming production. In an attempt to generate some warm Christmas feelings between himself and his son, Philip jokingly recounted how he had joined his high school drama club because he wanted to be around a girl who played the lead in the annual musical production. "I tried out for some bit parts but instead the club put me in charge of parking!" William only grimaced at this story and walked away. He told Philip that he was tired of his father trying to encourage him.

Philip was behind in buying presents and hated pushing through the crowds in jammed stores so he fell back on Internet shopping for the first time, which was convenient but depressed him. He vowed that, in the future, he would prepare earlier and buy nice things for Emily on his frequent trips. He used to bring her jewelry from all around the world back in the days when a Navajo brooch from Tucson airport was still an event.

Chapter Fifty-Four

The holiday season parties had begun soon after Thankgiving, and he was sorry to have missed one that featured carol singing at the dramatic seaside home of musical friends on the North Shore. But his favorite was held on Christmas Eve by friends who lived next to their church in Belmont. They had a wonderful recipe for warm scallop dip and for years he and little William used to compete for who could extricate himself soonest from all the hee-hawing after Christmas Eve service and race over to the steaming tureen to begin partaking of the crackers and heavenly broth. "There," Philip liked to joke, "was true salvation to be found."

It was hard to find any peace amongst all this holiday hubbub, but on the morning before Christmas Eve, while the kids were still sleeping in, he was able to catch up over breakfast with Emily. He remembered that she had gone out on the town with Maika and he asked how that went. "Very well," she answered, "I found out about her background finally, we usually just talk about our kids."

"That's nice," said Philip absent mindedly while perusing the headlines in the *Boston Globe* and sipping his coffee. "What did you find out?" "Well," said Emily, measuring her words a bit more carefully now, "for one, she told me that she has a much younger half-sister who is married to a French aristocrat and lives in a château in Normandy."

"Really," Philip answered, pricking up his ears and coughing drily. "I thought she was Czech." "She is," continued Emily, "her sister was a high fashion model with Lanvin and met her husband in Paris." Philip's mind started to look for cover in case the conversation went in an awkward direction. "In fact, Maika told me that her sister had mentioned last year that they had just hosted a meeting of astronomers at the château and she had met a charming professor named Philip from Harvard. Weren't you in Normandy at that time?"

"Heavens!" Philip exclaimed. "What a coincidence! I do remember her, the Countess who was a wonderful hostess for our event." Emily recalled her mother's admonition that often, in difficult moments, silence was best, so she said nothing and a troubling quiet stalked the room. But after a while she couldn't stand the tension any longer and asked Philip why he hadn't mentioned any of this at the time.

"I don't know, darling," he sputtered, "it didn't seem particularly important. Obviously I had no idea that there was a connection to anyone we knew here." "Obviously," huffed Emily, clearly upset. Philip bemoaned to himself the damper this revelation would put on Christmas cheer. Holidays seemed to somehow breed bad tidings.

Still attempting to put a positive spin on things, he remarked cheerfully that if Maika's sister ever came to Boston to visit, they could all meet. "No doubt we could," muttered Emily, not impressed by this attempt to save face. "It was awkward when Maika told me that her sister had mentioned you and I knew nothing about it. You must have made an impression." Philip apologized for the oversight but the damage was done and the incident cast a pall over their Christmas Eve that neither the minister's inveighing to forgive others, nor the scallop casserole afterwards, could entirely dispel.

Lying in bed that Christmas night Philip was troubled, wondering how it was that he, a faithful husband, had twice in the last couple of weeks found it necessary to dissimulate to his wife regarding matters concerning other women. Was this just a coincidence of past sins of his imagination crowding in on him suddenly? Did he need to change his behavior to avoid these awkward scenes with Emily? He didn't want to hurt her; somehow these tricky situations seemed to just find him without any effort on his part.

Rational man that he was, he was able to convince himself that Anna was history. She had told him herself that she probably wouldn't be allowed to keep in touch from her new identity. The Countess was discreet enough, he was sure, that the momentary connection that they had felt would stay with her, even if she did meet Emily at some future time.

Really, nothing had happened in either case except in his head, and surely that was his own business. We are entitled to have a fantasy world, aren't we, he thought to himself, even in this increasingly Orwellian world? Then, content in this reassuring thought, he slept soundly well into Christmas morn.

Upon awakening, it occurred to him that the friend that the Countess had recommended to him as a possible source of funding, a friend who was "interested in astronomy and very rich" must have been Ivan.

What a small world, he thought to himself, and how unwise to think that anything could remain private!

Christmas holidays quickly blended into the New Year and then it was time to start preparing for his course again and catching up with his two grad students. Besides his work on the proposed Institute, working with these young people was what Philip enjoyed most these days.

As he had predicted there hadn't been much feedback from the other three at the December meeting. They all assured him and Ivan that they had distributed copies of the proposal to key colleagues but had received little back. He set aside some time to identify some panelists who might assist him and Ivan to find a few candidates for faculty positions and planned to focus on that in January. He also suggested to Ivan that they have a separate meeting with the Divinity School Dean. "Just to get a fresh point of view; I think it would be helpful to us," he maintained.

Chapter Fifty-Five

Philip called the Divinity School Dean in early January to request a meeting and Ivan sent an advance copy of the proposal to inform him of the basic aims. They walked down from the Observatory together on a blustery January morning and were warmly greeted by the Dean's secretary and told that he would be with them shortly.

The Dean was a middle-aged, bespectacled, and jovial African-American man. He shook hands and after introductions welcomed them sincerely to the Divinity School. "I'll bet that you gentlemen have not passed through these doors before!" He joked, and they laughingly had to agree. "Please have a seat, and could I order you some coffee?"

After they had settled in he said that he had read the proposal and was impressed with Ivan's generosity in offering to provide funding for such a worthy cause. "I could not agree more with its aim to reduce the divisions in this country. No one of integrity could," he stated vehemently.

"The second aim, to gain insight into the phenomenon of faith and its roles in religion and science, sounds interesting," he continued, "but I'm less sure I really understand what you mean. I hope we can return to that in a moment.

"The third aim, to investigate whether the human mind is wired to grasp the beginnings and structure of the universe, also sounds fascinating, but it's not anything we in the Divinity School could contribute to, much less address the last issue you list, whether there may be a Darwinian disadvantage to having such wiring.

"Besides these aims that are explicitly called out in the proposal I seem to detect a further aim that requires some reading between the lines.

That is a desire to counter the opposition of the fundamentalist Christians in Congress to science funding and in particular to a certain large project requested by NASA. Am I reading that correctly?"

"There is some of that, you are right," Philip answered, "but it is in there largely to make it easier to obtain NASA support for the Institute. It shouldn't distract from our main aims that you have summarized very well."

"I see," said the Dean. "Then let me give you, if I might, some inputs on the material you sent me. My first thought is that if you are interested in helping to alleviate tensions in this country, you would do better to focus more of your effort on regions of the US where Christian fundamentalism is strongest and less on us here at Harvard.

"I could help you identify schools in the South where such an investment would be most helpful in showing folks that liberal intellectuals are trying to understand another point of view. There might be one or two historically black colleges in the mix, but I'd focus most on schools that draw heavily from the white fundamentalist population.

"Furthermore, I question whether founding an Institute at Harvard will make most efficient use of your resources. My advice would be to endow a few chairs in strategically chosen places of learning around the country and link them through support for workshops and creative use of electronic media. I could see a network of such Vesely Professorships attracting more of the kind of attention you are hoping for than creating yet another construct on our already over-burdened campus."

Ivan re-crossed his legs and ran his fingers through his hair, but didn't comment. He was seeing some advantages in what the Dean was saying. Philip was thinking that Cecily would probably have an easier time justifying NASA funds for some faculty positions at Christian fundamentalist schools than for a fancy institute at Harvard, and he also let the Dean go on.

"The role of faith in both fundamentalism and in modern science that you raise is certainly an interesting topic of study," the Dean continued. "But I wonder whether the approach you suggest, namely studying how educated liberals and conservatives manage to reconcile the Scripture with modern science, will bring the insight you are hoping for.

"As far as I can see, such individuals, and I include myself amongst them, have always seen the Scripture as a parable, not as literal history. Consequently they see no contradiction between biblical claims for the age of the Earth and the findings of archaeology, paleontology, or geology, for instance. They simply view the Bible's teachings as a valuable moral guide. To them, the battle to reconcile a literal interpretation of the Scriptures with science misses the point.

"So I don't think that you will gain much understanding of the nature of faith from studying such individuals. A more fruitful approach might be to ask if such a path towards reconciliation has always been freely adopted by many educated conservatives and liberals, why do one hundred forty million Americans still resist such an obvious reconciliation? Why do they dig in their heels and insist on the literal interpretation of the Bible? That's something that we in the church struggle with as much as anyone. Answering that question is where I'd put my money if you want to help science and, more broadly, help our country."

Philip and Ivan looked at one another and stood up to leave, thanking the Dean for his time. "You have certainly given us many things to think about," Ivan told the Dean. "We are grateful for your valuable insights; we'll discuss them with our colleagues and get back to you. We would certainly be most grateful for your assistance in identifying the most appropriate schools if we follow your suggestion on endowed chairs."

Philip agreed and they walked back towards the Observatory, treading carefully on the icy sidewalks up Mt. Auburn Street. The cold wind made it hard to talk so they mainly thought about what they had heard from the Dean. Ivan split off at the corner of his street, saying that Maika expected him for lunch at home. They agreed that Philip would summarize what they'd heard and distribute it to Cecily and the Science Dean and that he'd also drop in on Nancy to hear her reaction. They agreed soberly that this had been an important visit with a wise man.

Chapter Fifty-Six

When he got home, Ivan told Maika over lunch what the Divinity Dean had said. She agreed with him, also feeling that the Institute at Harvard would be an uphill battle probably wasteful of Ivan's energy and resources.

Over the next few days, much the same message came back from the others, in response to the notes on the meeting with the Divinity Dean that Philip had distributed. The Science Dean called Ivan and told him that the University would seriously consider his offer to endow one or more professorships in neuroscience, philosophy, or cosmology, but an Institute involving studies of faith would be hard to sell.

"We at Harvard are not known for pioneering new fields of knowledge," he said in a confidential tone. "We prefer to wait until the risky groundwork has been done elsewhere and then hire the key individuals, conferring upon them the prestige of our institution." At first, Ivan chuckled, thinking the Dean was being humorous, but when he heard no mirth on the other side, he realized that this was a serious statement of policy.

Ivan thanked the Dean and told him that he'd be back in touch. He was going to tell him that, in its beginnings, Harvard had once been more entrepreneurial when John Winthrop had asked Comenius to be the new College's first President in the early 1600s. But he thought better of it because the Dean probably wasn't all that aware of his University's early beginnings anyway.

Philip reported that his conversation with Nancy covered similar ground. She didn't want to lose Ivan as a big donor, she assured Philip, but she wanted to avoid a political quagmire. Instinctively she reverted to her reputation as a solid scientist administrator who "never made mistakes."

Cecily had also had gotten back to him confirming what he had expected, that the mood in Washington would be more supportive of initiatives at colleges in the fundamentalist heartland than in elite north-eastern universities.

Ivan and Philip agreed that they should take the Divinity School Dean up on his offer to help identify the schools that might be most appropriate for their initiative. The endowment of a chair or two at Harvard could be put in motion as soon as suitable candidates were identified by the panel. These first moves should put them on track to make a formal announcement of the program by the fall, they felt.

A few weeks later, Philip was surprised to get a call at his office from Anna saying that she was doing well in the Seattle area and working for Microsoft. She and Igor were no longer together because, after being questioned and settled in the mid-West by the FBI, Igor informed her that although he cared deeply for her, he was bisexual and had met a man who had promised to support him. He admitted that this was partly why he was anxious to leave Russia, where homosexuality was less accepted than in the US.

Anna also told Philip that she had gotten over this surprise and had since met a good man and they were talking of marriage. She didn't want Philip to worry about her. "I hope that you and your wife could attend our wedding when it happens," she added, sounding cheerful and happy. Philip was pleased that she sounded again like the Anna he had once known, the ingenue who had leapt into his arms when she brought the news that she had been awarded the NASA internship.

Philip pondered all this as he walked across the Observatory parking lot to his car that evening. Winter was loosening its grip and he hardly needed his coat. The approach of spring brought thoughts of renewal and of the passage of time. Mainly though, of the cycle of life. The kids would be back for Easter soon and it was time to start giving the spring segment of his course again.

He remembered how Anna had entered his life three years ago when she first sat in the front row of the class and smiled at him. Now she had met a "good man," whatever that meant, and wanted him and Emily to attend her wedding. He hoped that he wouldn't find a new pretty face smiling at him from the front row in his class this spring. Young people like her could get under your skin and it was exhausting. He preferred to always keep the memory of Anna as someone who had been special to him.

Chapter Fifty-Seven

Even with the best of intentions matters of importance take time, so it wasn't until a good part of a year later that invitations could be sent out to announce the inauguration of the initial Vesely-NASA Professorships at Harvard and at Emory and Brigham Young Universities. Cecily arranged for the ceremony to be held at the National Air and Space Museum on the Mall in Washington during cherry blossom time and the setting was splendid.

Attendees included the Harvard Deans of Science and Divinity, their equivalents at the two other universities, and the NASA Administrator. Several Congressmen from both sides of the aisle attended or sent representatives. It was a bipartisan event that made everyone feel good and their aides knew that there would be coverage in the *Washington Post.*

Other guests included the Count and Countess de M. from France, Dalibor and Zdenka from Prague, and the recently wed Anna Tsing and her delightful husband Mao, a talented microchip designer. Ivan and Maika were gracious hosts, and he gave a short speech stating the aims of the program, thanking all those who had contributed and introducing Professor Philip Frobisher as its Executive Director.

Philip and Emily hosted a lunch afterwards at the elite Cosmo Club whose membership is reserved for the most influential members of the US science and arts community. Cecily attended and brought her mum who wore her most outrageous Easter bonnet just like the one that Margaret Thatcher had favored.

At the lunch, Philip proposed a toast to the lovely person who put all this into motion. Asking the Countess to join him on the podium, he explained how, at a dinner in her château a few years ago, she had asked him what he had done that was important. "I was dumbfounded by this very simple question that all of us in science should be asked from time to time," he said, to cheers and laughter, "and I vowed that I would strive to do something that would provide her with a satisfying answer. Years later, I am very pleased that she has agreed to join us to see that her question put wheels in motion. Her brother-in-law Ivan Vesely and I hope that she agrees that the result is, indeed, important."

The press crowded forward and cameras clicked as the Countess signaled her approval by giving Philip a big hug and a kiss. Emily wasn't thrilled but she sat with the Count and Countess at lunch and admitted that she was delightful. "I'd never have believed that anyone as glamorous as she is could be so down-to-earth," she confided to Philip afterwards.

The program attracted favorable attention from both sides of the aisle. When a de-scoped version of COST began to look like a viable future mission, many pointed to the progress towards reconciliation that had been achieved as an important factor in its support from all quarters of the US political spectrum.

Ivan was pleased to see from the annual reports of the program's achievements that the role of faith in religion and science was being clarified. Most contributors to the program agreed that if faith in a benevolent Creator was shared by the majority of people on our planet, such faith may be a fundamental feature of the human mind. In that case, cosmological findings that threaten this belief might well carry a Darwinian disadvantage.

Discovery, for instance, that the multiple universe explanation of fine tuning of the universe fit the evidence better than the Divine Creator explanation could have negative social consequences. Many psychologists and sociologists agreed that casting billions of people spiritually adrift could be dangerous for the future of mankind.

Philosophers involved in the program pointed out that while Plato had already achieved some progress in defining Truth and Justice, he found Faith to be more difficult. It seemed to belong more to the world of myth than to his set of rationally accessible Forms.

Philip was interested to learn also that Plato had held up the beauty of ideas as a criterion by which our definitions of Truth and Justice, for example, are to be judged. He took satisfaction in learning that it was the beauty of the most perfect human forms that guided us towards the concept of Beauty as an abstraction. This wisdom of antiquity cast his affection for Estelle and Anna in a broader context that he found comforting.

Philip also found satisfaction in a final turn of events. He and Emily were hiking in Idaho and met the hunting guide who had been with Jock White on his final trip. He added a detail that hadn't been picked up by the media: "Ah reckon that cat had the greenest eyes I ever see'd," he told them. "Yep, that cat did!"

About the Author

Peter Foukal was born in Prague, Czechoslovakia, in 1945. He is a founder of CRI Inc., an electro-optics firm acquired in 2011 by PerkinElmer. He holds a B.Sc. from McGill University and a Ph.D. in Astrophysics from Manchester University, UK. He has taught and conducted research at the California Institute of Technology and Harvard and has served on advisory boards of the National Science Foundation, NASA, and the US National Academy of Sciences. He is a past-President of Division II of the International Astronomical Union and has served on the Corporation of Wheelock College, Boston. His 120 scientific publications include cover articles in *Science*, *Nature*, and the *Scientific American*. The 3rd edition of his text, *Solar Astrophysics*, was published in 2013. He and his wife live by the sea near Boston.